101 QUANTUM QUESTIONS

101 QUANTUM QUESTIONS

What You Need to Know
about the World You Can't See

Kenneth W. Ford

Harvard University Press
Cambridge, Massachusetts
London, England

Drawings by Paul G. Hewitt

First Harvard University Press paperback edition, 2012

Library of Congress Cataloging-in-Publication Data
Ford, Kenneth William, 1926–
101 quantum questions : what you need to know about the world you can't see /
by Kenneth W. Ford.
p.   cm.
Includes bibliographical references and index.
ISBN 978-0-674-05099-0 (cloth : alk. paper)
ISBN 978-0-674-06607-6 (pbk.)
1. Quantum theory—Miscellanea.   2. Quantum theory—Popular works.
I. Title.   II. Title: One hundred one quantum questions.   III. Title: One
hundred and one quantum questions.
QC174.13.F67   2011
530.12—dc22   2010034791

To
Joanne
and
Paul, Sarah, Nina, Caroline, Adam, Jason, and Ian
and
Charlie, Thomas, Nate, Jasper, Colin, Hannah, Masha, Ana,
Daniel, Casey, Toby, Isaiah, Naima, and Steven;
and
all the students it has been my privilege to teach,
every one of whom helped to enrich my life

Contents

IV   Quantum Lumps and Quantum Jumps

V   Atoms and Nuclei

VI   And More about Nuclei

Introduction

Big Ideas of Quantum Physics

There is no definitive list of "big ideas" in quantum physics. But here are a dozen that capture much of the essence of what quantum physics brought to the description of nature. What they all have in common is that they are inconsistent with "common sense"—that is, with how we expect the physical world to behave based on everyday experience.

There is a simple reason for the disconnect between common sense and quantum sense. The world we inhabit is a world in which quantum effects—and relativistic effects—do not directly impinge on our senses. We build our view of the world (at least of the physical world) on what we see, hear, taste, smell, and touch. It might have turned out that this world view continued to be valid in the quantum world—the world of the very small and the very fast—but it did not. So we are confronted with new ideas that seem weird and wonderful. They provide the threads running through this book. Be on the lookout for them.

You can imagine aliens who live out their everyday lives in the quantum domain. To them the ideas listed below are boringly obvious. To us Earthlings they are startling and mind stretching.

> **Quantization:** Nature is granular, or lumpy—both in the bits of matter that make up the world and also in changes that occur.
> **Probability:** Probability rules events in the small-scale world—even when we know everything there is to know about the events.

Wave-particle duality: Matter can exhibit both wave and particle properties.

Uncertainty principle: There is a fundamental limit in nature in the precision to which certain measurements can be made.

Annihilation and creation: All interactions involve annihilation and creation of particles.

Spin: Even "point particles" (those with no apparent physical extension) can spin, and spin is a quantized property.

Superposition: A particle or system of particles can exist in two or more states of motion at the same time.

Antisocial particles: Particles called *fermions* obey an exclusion principle. No two identical ones can occupy the same state of motion at the same time. Because of the exclusion principle, the periodic table exists.

Social particles: Particles called *bosons* can (and even "like to") cluster into the same state of motion, making possible an ultimate "togetherness" called the *Bose-Einstein condensation*.

Conservation: Certain quantities remain constant during all processes of change. Other quantities ("partially conserved quantities") remain constant during particular kinds of change.

Speed limit: The speed of light sets a speed limit in nature. (This consequence of relativity theory is most dramatically revealed in the quantum world.)

$E = mc^2$: Mass and energy are united into a single concept, so that mass can be changed to energy and energy to mass (another consequence of relativity theory that shows itself tellingly in the quantum world).

section I

The Subatomic World

1. What is a quantum, anyway? A quantum is a lump, a bundle. In the everyday world around us, there are lots of things that come in "lumps" of a certain size: loaves of bread, quarts of milk, automobiles. But there is no law of nature saying how big the loaf of bread or the bottle of milk or the automobile has to be. The baker could add or subtract a slice or even a crumb (see Figure 1). The dairy could decide to sell milk by the half-liter or by the pound. The car company could make its product just

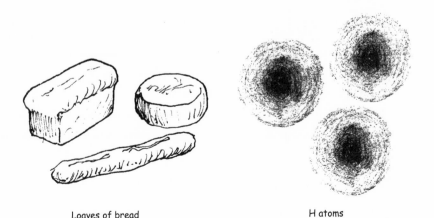

Loaves of bread   H atoms

FIGURE 1 Bread can come in any size. Ground-state hydrogen atoms are all the same.

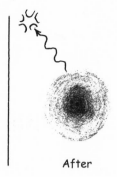

FIGURE 2
A hydrogen atom
in an excited state
emits a photon and
becomes a ground-
state hydrogen atom. .   **Before**   **After**

a little larger or smaller, just a little heavier or a little lighter. Not so in the small-scale world where quantum rules govern what happens.*

A hydrogen atom, to give an example, has a certain diameter (without a sharp edge, but that's a *different* quantum effect). Its diameter is about a tenth of a nanometer (a nanometer, or nm, is a billionth of a meter, or 10^{-9} m). It can't be made any smaller than that. That is its so-called ground-state size. And associated with that size is a certain energy, its ground-state energy. It can't be made to have any less energy than that. That is the basic hydrogen-atom lump. And *every* hydrogen atom in its ground state has exactly the same size and exactly the same energy— a universal hydrogen lump. The atom can be made larger and more energetic, but only in certain quantum increments, not by any chosen margin. These higher energy states, larger in size, are called *excited states* (see Figure 2). When an excited atom emits a photon of light and changes its energy from a higher to a lower value, it is making a quantum jump.

The photon that comes out is itself a quantum, a "lump" of light. It carries away a quantum of energy, and when it is absorbed—say by your retina—it deposits that quantum of energy at the point where it ends its life.

As the photon demonstrates, it is not just *things* that come in quantum lumps, but also certain *properties* of things. Energy, for instance.

Quantum can be an adjective as well as a noun. We speak of quantum physics and quantum jumps as well as a quantum of energy or a quantum of light. The plural of the noun is *quanta.*

And electric charge. Charge can't be trimmed finer than the amount carried by one proton (or its negative equivalent, carried by one electron). Nor can there be 3.7 quantum units of charge. Every charge in the universe is a whole-number multiple of the proton or electron charge. (There *is* an exception to this rule. Deep within fundamental particles, quarks carry charges of $1/3$ and $2/3$ unit, but they always combine in such a way that any observed or measured charge is a whole number of proton or electron charges.)

To give just one more example of a "lumpy" property (more examples will come later), the spin of every particle and every combination of particles is either zero or a whole-number multiple of a smallest spin, which is the spin of an electron. Loosely speaking, spin is a measure of the strength of rotational motion. Technically, it is a measure of a quantity called *angular momentum*. A child's top can spin. So can a merry-go-round. And so does Earth, as it turns once around its axis each day. The smallest amount of spin other than zero is the spin of an electron. This minimum quantum amount is possessed also by the proton. The photon, interestingly, has twice the spin of an electron or a proton. When you or I turn around, no matter how slowly, our spin is an astronomically large number of quantum units, so large that there is no hope of ever detecting the quantum nature of a person's spin.

For historical reasons, the adopted unit of spin is the same as that of the photon. It is said to have 1 unit of spin. The electron's spin, in this unit, is $1/2$. Then every particle and every entity in the world has a spin of zero or $1/2$ or 1 or $3/2$ or 2 or $5/2$, and so forth, never anything in between, never anything other than an integral or half-odd-integral multiple of the basic unit.

2. Where do the laws of quantum physics hold sway? The simple answer is "Everywhere." The real question is "Where do we have to pay attention to quantum physics?" To that the answer is "In the world of the very small, the world of molecules and still smaller atoms and still smaller atomic nuclei and still smaller fundamental particles—a world broadly described as the subatomic world (although it includes atoms, too)." It is in that small-scale realm where the lumpiness becomes important.

Imagine yourself walking over a gravel bed, over a sandy beach, and over soft, squishy mud. All of those surfaces are granular—pieces of gravel, grains of sand, molecules of mud. On the gravel bed you must step warily. You are very aware of the lumpiness. On the sand, you may be aware of the graininess, but it is barely significant. In the mud you have no awareness at all of the molecular "lumpiness." It is far below your ability to notice.

Or take a sip of water. You know that it is composed of H_2O molecules, yet that molecular lumpiness is irrelevant to you—or, for that matter, to the physicist analyzing the flow of water, its pressure, its viscosity, its turbulence, and so on. What is called *classical physics* (prequantum physics) is perfectly adequate to deal with water by the glassful or the pipeful or the reservoirful. But consider that single H_2O molecule. Classical physics can't handle it at all. It is a quantum entity and can be studied and understood only with the help of quantum physics.

So it's a matter of scale. Quantum physics is no less valid in our large-scale world than in the subatomic world. Indeed many of the properties of bulk matter can be understood only in terms of the quantum properties of its atomic and molecular constituents—things like how well it conducts electricity, how much heat is needed to raise its temperature 1 degree, what color it is. Yet the quantum underpinnings of such large-scale behavior hide from direct observation. Only in the atomic and subatomic realms does quantum physics leap out at us.

There are a few exceptions to this large-scale/small-scale dichotomy, a few instances where quantum effects make themselves felt directly in our everyday world. One of the most dramatic of these is superconductivity. In some materials at very low temperature, electrons move without resistance, literally without friction. This is a quantum effect reaching up to our human-sized domain. The frictionless motion of electrons is actually commonplace in the much smaller domain of atoms and molecules. There, electrons exhibit perpetual motion. They can move tirelessly forever, thanks to quantum rules that prevent their energy from falling below a minimum value. In a piece of wire, on the other hand, electrons normally experience resistance and stop flowing if not pushed along by an outside force—just as everything else in our ordinary world comes to rest if not propelled. So when you see a big superconducting donut in which electrons circle endlessly without being pushed, it means that the frictionless motion allowed by quantum physics has extended itself from the subatomic to the macroscopic domain. Electrons are literally circling in that big donut, just as they might circle within a single atom.

3. What is the correspondence principle? If quantum effects are important in the small-scale world and usually unimportant in the large-scale world, where is the dividing line? Put differently, where is the boundary between quantum physics and classical physics? That is actually a deep question that some physicists still struggle with. Some say, "Well, yes, quantum physics is valid everywhere and governs what matter does everywhere, but for large systems, special quantum effects (such as lumpiness) are irrelevant." Others say, "When we measure something that is going on in the subatomic world, we do so with large instruments in our classical world; the act of measurement causes the two worlds to be inextricably connected." Both agree that classical physics successfully describes nearly everything in the large-scale world and that quantum physics is needed in the small-scale world. They also agree that there is no conflict in the two ways of looking at nature, since quantum physics, whether evident or not, is always at work in every domain.

Let me take a moment to say what classical physics is. It is principally the physics developed in the seventeenth, eighteenth, and nineteenth centuries, the physics of force and motion (mechanics); of heat and entropy and bulk matter (thermodynamics); of electricity and magnetism and light (electromagnetism). Albert Einstein's twentieth-century theories of relativity—special and general—are also called *classical* because they are nonquantum. They deal not with quantum lumps and quantum jumps, but with smooth change. The classical theories are stunningly successful in the realms where they apply, be it powering a steam engine or broadcasting a radio signal or landing an astronaut on the Moon.

The great Danish physicist Niels Bohr was one of the first to grapple with the seeming conundrum that quantum physics, although totally different from classical physics, does not overturn it. In 1913 he introduced what he called the *correspondence principle*, which states that as the increments between one quantum state and the next become relatively smaller and smaller, classical physics becomes more and more accurate. In the hydrogen atom, for example, where Bohr first applied the principle, the ground state and the first few excited states differ dramatically from one another, and there is no resemblance to classical behavior, but if one goes to the one-hundredth or two-hundredth excited state, the quantum and classical descriptions begin to "correspond." It becomes possible to speak of an "orbit" for the electron, just as if it were a planet, and the quantum jumps of the electron from state 200 to state 199 to state 198 to state 197 match up well to a smooth inward spiraling, with radiation being emitted just as would be predicted classically. In fact, Bohr didn't just state the correspondence principle, he used it. Through the requirement that the quantum and classical worlds join smoothly for highly excited states, he was able to reach conclusions about the properties of all the states, right down to the ground state.

There is, in effect, a correspondence principle for relativity, too. Where gravitational fields are weak (by comparison with those near a black hole) and where speeds are small (by comparison with the speed of light), classical, nonrelativistic physics does just fine. Just as with quantum physics, however,

Niels Bohr (1885–1962). The young Bohr, pictured here, was better known to many Danes as a soccer (football) player than as a physicist. In his later years Bohr reportedly explained to a friend why he had a horseshoe mounted over the door of his summer home: "Of course I am not superstitious. I don't believe that a horseshoe brings good luck. But I'm told it works even if you don't believe."
Niels Bohr Archive, Copenhagen; Photo courtesy of AIP Emilio Segrè Visual Archives.

relativity doesn't cease to be correct for weak fields and slow speeds. Rather its effects are so small that for most purposes they can be ignored.

4. How big is an atom? If you are a proton, an atom is very big, about a hundred thousand times larger than you. If you are a person, very small, about 10 billion times smaller than you. So it is all relative. A single atom is too small to be seen with even the best optical microscope, yet it is enormously larger than many things now studied by physicists. It exists toward the upper end of the subatomic world. As mentioned above, the diameter of a hydrogen atom (in its ground state) is about a tenth of a nanometer, and heavier atoms are not much larger. A hundred thousand protons, if they could be lined up in a row, would stretch only across one atom. A hundred thousand atoms lined up wouldn't quite reach across the thickness of a piece of tissue paper.

5. What is inside an atom? Within an atom are electrons, protons, neutrons, and, depending on how you look at it, a lot of empty space. Before the twentieth century, some scientists doubted the existence of atoms, since the evidence for them was at best indirect. In 1905, when Einstein analyzed what is called *Brownian motion*, the existence of atoms became indisputable. Brownian motion (named after the nineteenth-century Scottish botanist Robert Brown, who first drew attention to it) is the incessant random motion of tiny particles suspended in a liquid. Brown had observed the motion for grains of pollen, and he hypothesized that the grains were moving about because they were living things. Einstein analyzed the motion—by then known to be exhibited also by inanimate particles—and showed that it could be fully understood by assuming that the particles (visible in a microscope) were being randomly bombarded on all sides by atoms or molecules. In other words, the invisible atoms and molecules were making their presence felt by pushing and jiggling larger particles that could be seen. Einstein's analysis indeed made it possible to estimate rather accurately the size and mass of individual atoms.

Yet for another half dozen years, the interior of an atom remained mostly mysterious. There was good reason to believe that the electron, which had been discovered in 1897, found a home within atoms. It was known to be much smaller and much less massive than a single atom, so logically it could be regarded as a constituent of atoms. Moreover, scientists took the fact that atoms emit light to be evidence for vibrating electrons within atoms. And, since the atom overall is electrically neutral (except when it occasionally gains or loses a bit of charge and becomes an ion) and the electron is negatively charged, there must also be positive charge within the atom—an amount of positive charge just sufficient to balance the negative charge of the electrons. What the nature of this positive charge might be or how it might be distributed within the atom was unknown at the time. (Unknown but not unguessed. Physicists like models that they can visualize, and some, in those early years of the twentieth century, espoused the "plum pudding" model, in which positive charge was spread over the whole interior of the atom like pudding and electrons were imbedded like raisins in the pudding.)

Ernest Rutherford (1871–1937). This 1907 pastel portrait by R. G. Matthews shows Rutherford in his lab at McGill University in Canada, where he conducted the work on radioactivity that won him the 1908 Nobel Prize in Chemistry. Today Rutherford's picture appears on the $100 bill of his native New Zealand. Mcgill University. Image courtesy of AIP Emilio Segrè Visual Archives, *Physics Today* Collection.

Clarity came in 1911 in the laboratory of an outsized, bluff New Zealander named Ernest Rutherford, working at that time at Manchester University in England, where he now and then gave loud voice to a hymn when things were going well. Rutherford and his assistants fired alpha particles from a radioactive source at a thin gold foil to study what small deflections the alpha particles might undergo in passing through the foil. Rutherford knew that alpha particles had a positive charge of 2 units and were ions of helium. He had no idea how small they were, but that did not matter for his calculations. He estimated (quite correctly) that if positive charge were spread uniformly within each gold atom, the alpha particles would emerge from the other side of the foil with a range of tiny deflections. To his astonishment, he found that a certain small fraction of the alpha particles underwent large deflections, some of them bouncing back toward the source. Rutherford said later, "It was quite the most incredible event that ever happened to me in my life. It was almost as incredible as if you fired a 15-inch shell at a piece of tissue paper and it came back and hit you."

It didn't take Rutherford long to figure out what was happening (see Figure 3). The atom's positive charge, he realized, must be concentrated in a small region—a nucleus—at the center of the atom. Then an alpha particle that penetrated the atom and got close to that nucleus would

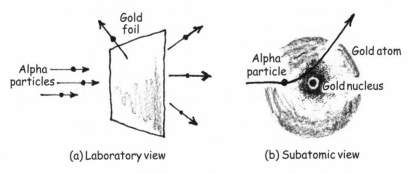

(a) Laboratory view   (b) Subatomic view

FIGURE 3   Rutherford's experiment with alpha particles aimed at a gold foil.
(a) Laboratory view. (b) Subatomic view.

experience a very large repulsive force and could be batted back.* Rutherford went on to analyze mathematically just what fraction of the incoming alpha particles should emerge in each direction, and the experimental results fit perfectly with the calculations. He couldn't say how large this central nucleus was, only that it had to be at least thousands of times smaller than the atom as a whole. Later experiments with more energetic particles were needed to reveal the actual size of the nucleus.

To visualize what is going on, first think of chopping your hand into a bowl of pudding. The direction of the motion of your hand might be altered ever so slightly, but basically it would just keep going. Then imagine that the pudding is put into some sci-fi compressor that turned it into extremely tiny, extremely dense nuggets of kryptonite, each one more massive than your hand. When you chop your hand through the collection of nuggets, you might miss them all, but if you encounter one, ouch! You would feel a large force that could make your hand rebound.

Now, about the empty space: There are actually two ways to look at the interior of an atom. One way is to visualize tiny electrons whizzing around in otherwise empty space, much as planets and asteroids orbit the Sun. It is perfectly correct to say that the solar system is mostly empty space, since the volume occupied by the Sun and all the planets, moons, and asteroids add up to far less than the whole volume out to Pluto or beyond. But electrons are not planets. Electrons spread in probability waves throughout the interior of the atom. A signature feature of quantum physics is the wave-particle duality (more on that later). According to wave-particle duality, an electron can be detected as a particle at a particular point in space—with a certain probability—but if you measure something else, say its energy instead of its position, it behaves like a wave spread out over the interior of the atom.

The ideas introduced here—waves, particles, probability—are core features of quantum physics and will be themes throughout this book.

---

*Actually, a large attractive force would have nearly the same effect, but Rutherford knew that attraction by tiny lightweight electrons couldn't be responsible for the large alpha-particle deflections, so the nucleus had to be positively charged.

6. Why is solid matter solid if it is mostly empty space? Think of a spinning airplane propeller. If you look at it with a strobe light, you see that it occupies only a small part of a disk. But if you take a time exposure, it seems to fully occupy a whole disk. Back in World War I, fighter pilots fired machine gun bullets through their whirling propellers with no ill effect. But if they had tossed a baseball toward the propeller instead, it would have been chewed to pieces or batted back. Electrons in atoms are the same. A fast alpha particle passes right through the electron cloud as if it weren't there. A big, slow atom, on the other hand, approaching a neighboring atom gets kicked back as if it were encountering a solid sphere.

If you lean your elbow on a table, your atoms and the table's atoms come face to face. The whirling electrons (or electron clouds) in each atom prevent penetration by the other atom. They act like solid balls. Actually, because atoms have no exact skin or edge, physicists prefer to talk in terms of the forces that atoms exert on each other rather than about what happens when they "touch." When there is a small distance between them, atoms attract one another, but at a certain point as they get closer, they repel, and when they get quite close, the repulsive force becomes very large, preventing the two atoms from significantly mingling.

As happens so often in the quantum world, what result you get depends on what experiment you do—on what you measure. If, hypothetically, you rolled one atom toward another, it would bounce away, revealing the other atom to be, apparently, a solid ball of a certain size (see Figure 4). What happens is about the same as what would happen if you rolled one bowling ball toward another. You would conclude that matter is solid and you would understand why your elbow doesn't sink into a table. But if instead you fired a "bullet" (such as an alpha particle or an electron) at the atom, the high-energy particle would, more likely than not, go right through. You might conclude that matter is mostly empty space.

Let me now jump right to a fundamental question of quantum physics: Is the electron in an atom a particle or is it a wave? The usual answer is that it is both. It is a particle that spreads itself over a region of space in a probability wave. It can show itself as an amorphous cloud or as a

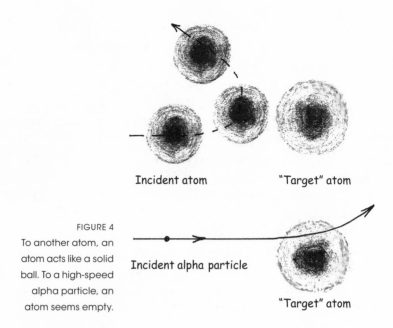

Incident atom   "Target" atom

FIGURE 4
To another atom, an
atom acts like a solid
ball. To a high-speed
alpha particle, an
atom seems empty.

Incident alpha particle

"Target" atom

particle, depending on how you look at it. I prefer to say that it really *is* a particle, for it can be created at a point, annihilated at a point, and detected at a point, but that it can *behave* as a wave, which it does when occupying some state of motion in an atom or propagating from one place to another.

This wave-particle duality is difficult—perhaps impossible—to visualize. It conflicts with common sense—our expectation of the way things should be based on everyday experience. Common sense tells us one thing; quantum physics may tell us something different. This can be unsettling, but it shouldn't be too surprising, for our common sense is based on our experience in the classical world. We learn about the quantum world not directly through our senses but indirectly through a set of measuring instruments. Perhaps quantum physics might have turned out to be consistent with common sense, but it didn't. We just have to get used to that fact. Unsettling, yes, but exciting, too.

7. How big is an electron? Is there anything inside it? The electron, the first fundamental particle to be discovered, made its maiden appear-

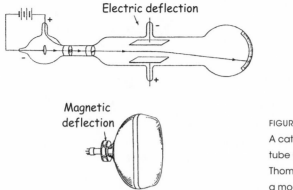

Electric deflection

Magnetic
deflection

FIGURE 5
A cathode ray
tube used by J. J.
Thomson (top) and
a modern CRT.

ance in something called a *cathode ray tube* in the laboratory of J. J. Thomson at Cambridge University in 1897.* He and other scientists at the time knew that if a glass tube from which most of the air had been removed contained within it two metal plates, one at each end, with one charged positively and the other charged negatively, some kind of "ray" flowed from the negative to the positive plate. Since the negative plate was called the *cathode* (with the positive plate being called the *anode*), these unknown rays were called *cathode rays*. Thomson set about trying to find the nature of these rays. He built tubes of various kinds (one of them is shown in Figure 5) containing various gases at various pressures, and he deflected the rays with both magnetic and electric fields.† From these careful studies, he reached several conclusions. The rays are in fact negatively charged particles. Because they pass readily through dilute

---

*Thirty years later, J. J.'s son George, who was a child of five when his father discovered the electron, was one of those who showed that the electron has wave properties. Both J. J. and George won Nobel Prizes for their efforts.

†As it happens, we use cathode ray tubes, or CRTs, today in much the same way that Thomson used them. In the CRTs that were ubiquitous in television sets and computers before flat-screen displays took over, electrons are accelerated to high speed by an electric field and then deflected toward chosen points on the screen by a magnetic field. In a modern oscilloscope, whose principles of construction exactly mirror those in the Thomson CRT shown in Figure 5, electric fields are used for both acceleration and deflection of the electrons.

gases, they must be small in size. And because they are deflected in certain measurable ways by magnetic and electric fields, they must have a ratio of mass to charge *(m/e)* that is small. By *small* he meant small relative to the mass-to-charge ratio of a hydrogen ion (which, as we now know, is just a proton)—roughly a thousand times smaller, he reckoned.

Concerning this latter finding, Thomson wrote, with care, "The smallness of *m/e* may be due to the smallness of *m* or the largeness of *e*, or a combination of these two." As we now know, it is due to the smallness of *m*. In fact, the magnitude of the electron's charge *e* is exactly the same as that of the proton, whereas the mass of the electron is nearly two thousand times less than that of a proton.

Thomson didn't know the size of a single atom. But whatever an atom's size, he inferred that the electron is much smaller. Now we believe that the electron is as small as it is possible to be, a true point. It has no size. Certainly it has no known constituents. It is one of the particles we call *fundamental*, as opposed to *composite*. (The proton, by contrast, is composite.) How, you might ask, can a particle have mass and yet no size, charge and yet no size, spin and yet no size? How can it *exist* and have no size? The only answer is that quantum physics allows an object that exists at a mathematical point to be endowed with various physical properties. According to the extremely successful theory of electrons and photons that goes by the name *quantum electrodynamics (QED)*,* when an electron interacts with a photon, it does so at a point in space and time. In such an interaction, an electron is actually created or destroyed, and always at a point.

Yet quantum physics has more to say about the size of an electron. Because evanescent particles called *virtual particles* are constantly being created and destroyed, an electron travels with a retinue of companions—a constantly changing retinue, like a king strolling down a pathway accom-

---

* Richard Feynman has written a lovely and readable book called *QED: The Strange Theory of Light and Matter* (Princeton, N.J.: Princeton University Press, 1986). A play by Peter Parnell about this remarkable physicist is titled *QED*. (In Los Angeles and New York, Alan Alda gave a strikingly realistic performance as Feynman.)

panied by a changing cast of attendants, some of whom at any moment are joining the procession while others are leaving it, with the total number of attendants constantly being altered. If the king is never permitted to travel alone, his "size" is bigger than the size of his physical self. And, as I have already discussed, an electron in an atom is governed by a probability wave extending over the whole atom. It's as if the electron, in a particular state of motion in the atom, is spread over a big volume.

Yet, at the most fundamental level, the electron appears to be a dimensionless point. This conclusion, like every conclusion in physics, is tentative. All we can say for sure, based on experiment, is that if the electron has a size, it is at most thousands of times smaller than a proton, or hundreds of millions of times smaller than an atom. Present-day theory treats it successfully as a point. Lurking in the wings, however, and perhaps about ready to march onto the stage, is another theory called *string theory* (addressed in Question 99), which sets out to unite quantum physics and gravity. According to string theory, electrons and all other fundamental particles are not point entities but are bits of vibrating string. The size of these strings is small beyond imagining. A thousand million billion of them wouldn't be enough to stretch across a single proton.

In all this discussion, I should not neglect to mention the practical value of what we might reasonably call "our friend, the electron." Whether on the surface of the Sun or within a compact fluorescent lamp, it is the motion of electrons that causes light to be emitted. In the retina of your eye, it is electrons within molecules that absorb light and let you see. In high-tension lines and motors and generators and computers and household gadgets, it is electrons that do the work of the industrial age. And within every living cell, it is the trading of electrons back and forth that powers life.

Digging Deeper

8. How big is a nucleus? What is inside it? Let me start with the second question. Inside the nucleus are protons and neutrons. The proton carries positive electric charge; the neutron is electrically neutral (thus its name). Although very different electrically, these particles have comparable mass and comparable size. The proton's mass is some 1,836 times that of the electron; the neutron is 1,839 times more massive than the electron. Both measure a bit more than 10^{-15} m across, a hundred thousand times smaller than an atom but some thousands of times larger than the smallest distances now being probed in particle physics experiments. They are so "big" because they are composite, not fundamental. Three quarks (which *are* fundamental), not to mention some gluons, swirl within each proton or neutron. However, for this discussion, I will ignore the quarks and the gluons. Much about nuclei can be understood just in terms of their proton and neutron constituents—particles collectively called *nucleons*.

The proton was, in essence, "discovered" through Rutherford's 1911 work showing that a nucleus exists at the center of every atom. The nucleus of the simplest atom, that of hydrogen, came to be called the *proton*. For two decades, physicists mostly assumed that nuclei were composed of protons and electrons—protons to give the needed mass and electrons to offset some of the charge of the protons. This model had numerous difficulties, among them that there was no known force that could bind

Enrico Fermi (1901–1954), shown here with his wife, Laura, and their children, Giulio and Nella, probably on their arrival in the United States in December 1938. Physicists like to say that Fermi got lost on his way from Sweden (where he had just received the Nobel Prize) back to his native Italy and arrived accidentally in America. Laura Fermi later wrote a wonderful book, *Atoms in the Family*. Photo courtesy of AIP Emilio Segrè Visual Archives, Wheeler Collection.

electrons so tightly to protons. Also, because of its wave nature, the electron would resist mightily being confined in so small a space.*

These difficulties were swept away in a single stroke when James Chadwick in Cambridge, England discovered the neutron in 1932. Suddenly it was clear: Nuclei are made of protons and neutrons. Yet the clarity

---

* The discussion at Question 65 will help explain why this is true.

wasn't quite unblemished. A puzzle remained: If there are no electrons in the nucleus, how can it be that electrons shoot out of nuclei in the process of beta decay? It remained for Enrico Fermi in Rome, two years later, to explain this. Fermi offered a theory in which electrons are created at the moment they are ejected from nuclei. This groundbreaking theory underlies everything that we have learned about particles since. *All* interactions of *all* particles involve creation and annihilation of particles. Our seemingly stable world is built on a near-infinitude of catastrophic events in the subatomic world.

Once the neutron's discovery allowed physicists to breathe a sigh of relief as they banished electrons from the nucleus, they had to ask whether nucleons in the nucleus are locked in fixed positions, like the sodium and chlorine ions in a salt crystal, or jostle about like H_2O molecules in a drop of water, or fly about like oxygen and nitrogen molecules in the air. The liquid droplet model, championed by Niels Bohr in the 1930s, seemed best able to explain what was then known about the properties of nuclei. In a real tour de force, Bohr and his younger colleague John Wheeler in Princeton used the liquid droplet model to account for the newly discovered phenomenon of nuclear fission in 1939. Yet a decade later, new information about the behavior of nuclei began to hint that the nucleons were, after all, free to move about within the nucleus like gas molecules. What then came to the fore and is still a pretty good description of nuclei is the so-called unified model, or collective model, in which the nucleus exhibits both liquid and gas properties. It is one of the oddities of quantum physics that protons and neutrons, although held as closely together as water molecules in a liquid, are still free to race about, through and among each other, like molecules in a dilute gas. It's as if guests at a very crowded cocktail party could miraculously shoot across the room straight toward a friend or the drink table.

Which leaves the question "How big is a nucleus?" It turns out that the volume of a nucleus is simply proportional to the number of nucleons in it. A nucleus containing one hundred nucleons, for instance, has twice the volume of a nucleus containing fifty nucleons and four times the volume of a nucleus containing twenty-five nucleons. But the *radius* of the nucleus increases only in proportion to the cube root of the volume, so doubling the volume increases the radius by only about 25 percent. Because of this relationship between radius and volume, nuclei of heavy elements are not vastly larger than nuclei of light elements. The nucleus of a uranium-238 atom, for instance, is only about six times larger than a single proton.

One last small fact about nuclei. They are not all spheres. Some are deformed into prolate (football) shape, and a few into oblate (pancake) shape. This results from the relatively free motion of some nucleons within the nucleus. If a nucleon orbits within the nucleus in such a way that its probability wave is not spherical, the nucleus as a whole tends to shift its shape to match that of the orbiting nucleon. Picture a blown-up

Robert Hofstadter (1915–1990), shown here at Stanford University in November 1961 following the announcement of his Nobel Prize. Hofstadter's move from Princeton to Stanford in 1950 was fortuitous, for it provided him with just the instrument he needed, a powerful linear accelerator. Not content only to do superb experiments, he also mastered the theory behind the experiments. Photo courtesy of Jose Mercado, Stanford News Service.

balloon inside which a gerbil is running round and round in a circle. The balloon will distort a bit to give the gerbil more running room.

9. How big are protons and neutrons? What is inside them? A proton is a ball of positive charge that spins and generates a magnetic field and measures roughly 10^{-15} m (a millionth of a billionth of a meter) across. A neutron is about the same size. It also spins and generates a magnetic field. Although it has no net charge, it is not totally neutral. Within it are balancing positive and negative charges.

The first major steps toward probing the interior of protons and neutrons were taken by Robert Hofstadter in the mid-1950s, using an electron accelerator at Stanford University. His method bore a striking similarity to that of Ernest Rutherford some forty-five years earlier. Rutherford learned about the interior of atoms by measuring the deflections of alpha particles fired at the atoms. Hofstadter learned about the interior of nucleons by measuring the deflection of electrons fired at nuclei. Rutherford's alpha particles were a few million electron volts (MeV) in energy.* Basically he learned just one thing in his early experiments: that the atom's positive charge is located in a tiny nucleus deep within the atom. Hofstadter, at first with 600-million-electron-volt (600 MeV) electrons and later with electrons of more than a billion electron volts (1 GeV), learned much more. He could measure not only the sizes of nuclei and of the nucleons within them, but also gain some knowledge of how electric charge and magnetism are distributed within them. Figure 6 shows graphs of the charge density within a proton and a neutron.

As I shall discuss later, every particle has a wavelength, and the greater the particle's energy, the less its wavelength. The less its wavelength, in turn, the better able it is to probe small regions of space. A limit to what an ordinary microscope can reveal is set by the wavelength

---

* An electron volt (1 eV) is the amount of energy gained by an electron in being accelerated through an electric potential difference of 1 volt. Common units in the nuclear and particles worlds are millions of eV (MeV) and billions of eV (GeV), sometimes even trillions of eV (TeV).

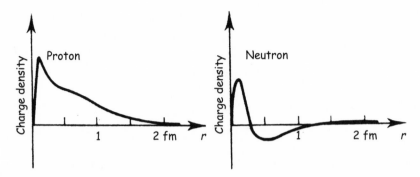

FIGURE 6 Graphs showing the density of electric charge within the proton and the neutron, as a function of distance from the particle's center. The density is presented per unit of radial distance rather than per unit of volume.

of light that it uses. An electron microscope does better because its electrons have wavelengths far less than the wavelengths of light. The Stanford electron accelerator was, in a sense, a very large and very powerful electron microscope. The wavelengths of its billion-volt electrons were less than (but not *much* less than) the diameter of a single proton, so they could illuminate that particle's interior.

Still eluding Hofstadter, however, were the quarks that we now know lurk within every proton and neutron. Quarks are fundamental particles, which, like electrons, have a spin of ½ unit and, also like electrons, seem to be point particles of no physical extension. I will deal more with quarks later in this book (see Question 45). Before their existence was confirmed experimentally, they were postulated (in 1964) independently by two young theorists, Murray Gell-Mann and George Zweig. It is to Gell-Mann that we owe the name *quark.*

Although there is a wealth of data now pointing to the reality of quarks, the fact is that no single one has ever been seen. They are maddeningly shy, wanting to cluster always in groups of two or three. The reason they can't be pulled apart is that the attractive force between them grows stronger as they get farther apart (quite unlike the gravitational or electric force). They thwart every effort to separate them by, if necessary, creating new quarks such that none is ever isolated. Here is a far-fetched analogy: In a science-fiction film, aliens Beta and Zeta are locked in an embrace. Earthlings Rob and Bob apply enormous force to pry them apart. Finally, after great effort, they succeed, or think they succeed. As Beta and Zeta come apart, a new Beta and new Zeta are created. The new Beta attaches itself to the old Zeta and the new Zeta attaches itself to the old Beta. After all of the huffing and puffing by Rob and Bob, they don't get a separated Beta and Zeta. Instead they get two Beta-Zeta combinations where there had been one.

10. What is Planck's constant and what is its significance? Max Planck, a professor in Berlin, ushered in the quantum era in 1900 (hardly knowing that he was starting a revolution). He was puzzling over what is called *cavity radiation,* which I shall explain in a moment. In that year classical physics reigned supreme, and physicists could be excused for feeling good—dare I even say smug—about understanding so much of the physical world with the help of the great theories of mechanics, thermodynamics, and electromagnetism. To be sure, there were a few small chinks in the edifice of classical physics, but most physicists assumed that classical physics itself could be used to repair those chinks. For instance, no one knew why atoms emitted the particular frequencies of light that they do. Nor did anyone understand the recently discovered phenomenon of radioactivity. Since no one knew how to attack these two problems, it was permissible to assume (or hope) that classical physics could be used to explain them. But there was another difficulty in the edifice of physics that was more troubling because physicists thought that they *should* be able to explain it. That was the mystery of cavity radiation.

Everything emits radiation, with an intensity and a range of frequencies dependent on its temperature. For the Sun or a campfire, the radiation is readily visible. For an electric heater element just starting to warm up, the red glow is barely visible. For the wall in your bedroom, the radiation is not visible, for it is weak and it is in the infrared. But it is as surely present as is sunlight at noon. Now replace the bedroom with a large box, its walls held at some temperature. The walls continually radiate inward and they also continually absorb radiation. Within the box— the cavity—electromagnetic waves of many different frequencies are running this way and that (Figure 7). What scientists had learned before

Max Planck (1858–1947), shown here c. 1908, the year he turned fifty. In 1914 he invited Einstein to Berlin, and the two became friends, playing music together (Planck on the piano, Einstein on the violin). Planck's older son, Karl, was killed in action in World War I, and his younger son, Erwin, was executed by the Gestapo in January 1945 after being implicated in the July 20, 1944 attempt on Hitler's life. Planck refused to join the Nazi party, even though doing so might have spared Erwin's life. Photo courtesy of AIP Emilio Segrè Visual Archives.

Planck's work was that the radiation within the cavity is determined solely by the temperature of the walls, not by the composition of the walls and not by the size of the cavity. How the intensity of that cavity radiation depends on frequency is shown in Figure 8.

Up until Planck's work, every effort to explain the distribution of intensities depicted in Figure 8 failed utterly. Classical theory even absurdly predicted infinite intensity of radiation in the cavity. In the late fall of 1900, after repeated efforts, the forty-two-year-old Planck did finally succeed in accounting for cavity radiation, but to do so he had to perform what he called "an act of desperation." He postulated that the "resonators" (vibrating electric charges) in the walls of the cavity emit and absorb radiation only in bundles that he called *quanta*. Moreover, each bundle—each quantum—in his theory carried an energy directly proportional to its frequency. He wrote the equation

$$E = hf,$$

in which E is the quantized bundle of energy emitted by a resonator, f is the frequency of the emitted radiation (say in vibrations per second, or hertz), and h is a new constant that he introduced. Quanta of high frequency have more energy than quanta of low frequency. Planck used the

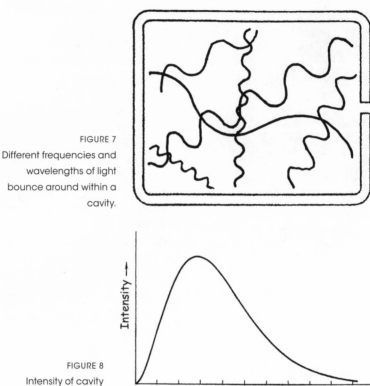

FIGURE 7
Different frequencies and wavelengths of light bounce around within a cavity.

FIGURE 8
Intensity of cavity radiation.

letter h to designate his new constant, and no one has been of a mind to change it since. Not surprisingly, h is known as *Planck's constant*. From known data on cavity radiation, Planck himself was able to calculate a reasonably accurate value of h. In the everyday units of joules for energy and hertz for frequency, it is *extremely* small, less than 10^{-33}.

Planck's constant ushered in a wholly new idea in physics, the idea that actions and change in the world around us are not smooth and continuous, as imagined in classical physics, but are granular, or "lumpy." Moreover, by its numerical magnitude it established the scale of the small-scale quantum world. We can imagine a hypothetical world where h is much greater than in our world. There quantum effects might be large and obvious. Or we can imagine a world in which h is much smaller than in our world.

My license plate proclaims my love for quantum physics. When asked what H BAR means, I sometimes say that it is the name of my cattle ranch in New Mexico. Anyone checking out this claim might find that there is an actual BAR H BAR (- H -) ranch in Idaho.

There, quantum effects would be restricted to even tinier domains that are even farther removed from the everyday world. Or we can even imagine a world in which h is magically reduced to zero. That would be a world without quantum effects, a purely classical world, a world in which there would be no need for books like this one. (It might also be a world in which cavity radiation had reduced everything to toast.)

One small postscript on Planck's constant. As quantum physics evolved in the early part of the twentieth century, physicists kept finding that h showed up in their equations divided by 2π. For instance, the strength of an electron's orbital motion in an atom (its so-called orbital angular momentum) was found to be equal to $h/2\pi$ or $2(h/2\pi)$, or $3(h/2\pi)$, and so on. So, as a shorthand, physicists starting writing $h/2\pi$ as $\hbar$, pronounced *h bar**.

11. What is a photon? Nearly everyone knows that Albert Einstein discovered the theories of relativity (special and general). Fewer people know that he also contributed to the birth of quantum physics. In 1905, the same year in which he proposed the theory of special relativity—thereby uniting space with time and mass with energy—Einstein argued that light is not only emitted and absorbed in bundles, as Planck had proposed, it *exists* in bundles. At first physicists called these quanta

---

* If you want to sound knowledgeable when visiting Germany, which was the epicenter of theoretical physics in the first third of the twentieth century, pronounce it *ha-strich*.

Albert Einstein (1879–1955), shown here c. 1920 in Berlin. So confident was Einstein that he would receive the Nobel Prize (which he did, in 1921) that as part of his divorce settlement with his wife Mileva in 1919, he stipulated that his Nobel award would be put into a trust fund for their two sons. In his later years in Princeton, Einstein played the part of the absent-minded professor with his unruly hair, baggy sweater, and unmatched socks. Photo courtesy of Niels Bohr Archive, Copenhagen.

of electromagnetic energy *corpuscles*. In 1926, Gilbert Lewis proposed the name *photon*, and it has stuck.

The "-on" at the end of the word *photon* suggests that it is a particle, just as electrons, protons, muons, and pions are particles. In truth, physicists didn't open the door and invite the photon in as a full-fledged particle until the early 1930s, when the successful theory of its interaction with electrons gave it complete legitimacy. At first physicists found it difficult to think of something with obvious wave properties as a particle.* Even when they had come to terms with the wave-particle, duality, they found it hard to imagine a particle with no mass. Now we regard the photon as no less a particle than the electron or neutrino or quark. Like other particles, it has energy, momentum, and spin and can be created and annihilated. Yet it does stand apart in one way. Because it has no mass, it travels always at the same speed. Other particles can be slowed or stopped. Not the photon. From the moment of its creation, and no matter what its energy, it moves always as fast as anything possibly can move, at nature's speed limit—called, of course, the *speed of light*.

12. What is the photoelectric effect? When ultraviolet (UV) light strikes a metal surface, electrons fly out of the surface. Evidently some of the energy in the UV light has been transferred to the electrons, which can then break free of the metal and fly away. This is called the *photoelectric effect*. The effect was known at the time of Einstein's 1905 work, but its details were not. It remained to determine how the number and energy of the electrons depend on the nature of the incoming light—in particular, on its intensity and frequency. Using his particle model of light, Einstein predicted these details exactly. It was for this achievement, not relativity, that Einstein was awarded the Nobel Prize sixteen years later.

Classically, one would expect that more intense light would result in more energetic electrons and that light of any frequency, if sufficiently

---

* When Planck supported the nomination of Einstein for membership in the prestigious Prussian Academy of Sciences in 1913, he stated: "That . . . in his hypothesis on light quanta, he may have gone astray in his speculations should not be held against him."

intense, would release electrons. Neither of these expectations is confirmed by experiment. Instead, the energy of the electrons depends only on the frequency of the light, not its intensity, and light below some cutoff frequency does not release any electrons at all, regardless of its intensity. These facts about the photoelectric effect were simply explained with Einstein's particle model of light. Visualize the incident beam of light as a stream of photons (I use the modern terminology). The energy of each photon is given by Planck's formula, $E = hf$. The photon's energy can be transferred in its entirety to an electron, but it cannot be sliced into smaller pieces or slowly dissipated as the photon slows down (it *can't* slow; it's all or nothing). So an electron, upon absorbing a photon, gains the photon's full energy. Some of that energy the electron needs to overcome an electric-field barrier at the metal surface. The rest is available as kinetic energy for the escaping electron. Below a certain frequency—a cutoff frequency—the energy transferred to an electron is less than it needs to overcome the barrier so that the electron won't escape. No matter how intense the light, no single electron will gain enough energy to break free. But if the light's frequency is above the cutoff value, the photons will provide electrons with enough energy to escape. The number that emerge will depend on the light intensity, but the energy of each one will depend only on the frequency.

Measurements by the American physicist Robert Millikan in 1916 brilliantly confirmed all of Einstein's predictions about the photoelectric effect. Five years later, Einstein was awarded the Nobel Prize "for his services to Theoretical Physics, and especially for his discovery of the law of the photoelectric effect." (Notice that the citation makes no mention of the quantum nature of light.) Then, in 1923, Millikan joined Einstein as a Nobelist "for his work on the elementary charge of electricity and on the photoelectric effect."

13. What particles are believed to be fundamental? What particles are composite? The definition of a fundamental particle is very simple: a particle not known to be made of other particles. Fundamental particles appear to have no physical extension and to interact at spacetime points. Composite particles, as their name suggests, are made of other particles. They do have a physical extension, and are created and annihilated over small regions of space. Electrons, photons, and muons are fundamental. Protons, neutrons, and pions are composite.

Physicists have identified twenty-three fundamental particles, not counting their antiparticles or the hypothetical carrier of the gravitational force, the graviton, and also not counting the still-to-be discovered Higgs particle (see Question 98). Hundreds of composite particles are known, and there is no limit to how many there might be (just as there is no limit to the number of chemical compounds that can be created from ninety-two elements).

Particles are perfect exemplars of quantum physics. When they interact with one another, when they emit and absorb light, when they lock into certain states of motion or flip from one state to another, when they morph from particle to wave, they are about as nonclassical as it is possible to be. So I will call on them frequently in this book to illustrate points about quantum physics.

Particle physics can be said to have been launched in 1897 when Thomson discovered the electron. At first the field moved at a stately pace. By 1930 the electron had been joined by only the proton and the photon. Then the pace quickened slightly. By 1940 the neutron and some mesons (later identified as *pion* and *muon*) had joined the party, and the neutrino had been surmised. In the late 1940s the field picked up speed with the arrival of more new particles, some of them called *strange*. The 1950s and 1960s brought a flood of new particles.

Physicists then rightly concluded that what they had been calling *elementary particles* could not all be elementary. Some were indeed fundamental; most were composite.

In Section VII you will find more about the particles and their properties. Here I present a quick overview of those that are now considered fundamental. First there are six particles called *leptons*: the electron, the heavier muon, the *much* heavier tau, and three neutrinos, one each of the electron, muon, and tau type.* The electron, muon, and tau carry a negative charge; the neutrinos are uncharged (neutral) and have only very wispy masses, much less than the mass of even the lightweight electron. What the six leptons have in common is that they do *not* experience the strong force that holds nuclei together. They experience the weak force and, if charged, the electromagnetic force. The leptons are summarized in Table A.1 (located, along with Tables A.2, A.3, and A.4, in Appendix A.)

Six down and seventeen to go.

Then come six particles that *do* experience the strong force. These are the quarks. Like the leptons, they have ½ unit of spin, and they also experience the weak and electromagnetic forces. Their properties are summarized in Table A.2. One of their oddities is that they carry an electric charge of magnitude ⅓ or ⅔ of what had previously been believed to the smallest unit of charge. But, as I mentioned earlier, because quarks team up always in twos and threes, no entity with fractional charge has ever been seen.

Physicists unimaginatively named the first two quarks to be discovered the *up* and *down quarks* (just names, nothing to do with direction). Then came the fancifully named *strange* and *charm quark* (yes, *charm*, not *charmed*). Then came the *top* and *bottom* quarks (I have always regretted that the quirky physicists who wanted to call them *truth* and *beauty* didn't get their way).

That brings us to twelve. The other eleven fundamental particles are called *force carriers* or sometimes *exchange particles*. Their properties (along with those of the hypothetical graviton) are summarized in Table A.3. One of them, the photon, accounts for the electromagnetic interaction. Two of them, called the W and Z particles, account for the weak interaction.† The remaining eight are the aptly named gluons that provide the nuclear glue of the strong interactions. Here is the stunning fact about the force carriers: Every interaction in the world is brought about

---

* Physicists have good reason to believe that these three "families" are all that there are. I touch on the reasons for this conclusion at Question 42.
† The Z is neutral and is usually designated Z^0. The negative and positive Ws (W^- and W^+) are each other's antiparticles.

by their exchange among leptons and quarks. And *exchange* means the annihilation and creation of the particles that are involved. The next time you take someone by the hand, think of the subatomic chaos that is required to make this contact possible.

14. What is the standard model? The array of twenty-three fundamental particles, the composites they form, the three interactions they generate, and one more still-theoretical particle called the *Higgs* constitute what has come to be called the *standard model*. Are its days numbered? No one knows. But if the string theorists someday prevail, the "new, improved" standard model will have fewer than twenty-three fundamental entities and will incorporate the fourth interaction, gravity, into the fold.

section III

The Small and the Swift

15. What are some quantum scales of distance? A nanometer is a billionth of a meter (written as 1 nm, or 10^{-9} m), and it is a handy unit in the atomic and molecular realm. Ten hydrogen atoms standing shoulder to shoulder would stretch about 1 nm, and four or five water molecules would cover about the same distance. Tiny circuits or structures of dimension 10 to 100 nm are said to be at the nanoscale. It is a scale right at the boundary line between the quantum and classical worlds. Some X-rays have a wavelength of about 1 nm. The light that you see has wavelengths about five hundred times greater (actually spanning 400 to 700 nm).

Many of the phenomena discussed in this book occur at dimensions well below the nanoscale. A handy unit in the nuclear realm is the femtometer, sometimes called the *fermi* (and abbreviated *fm* in either case). It is one millionth of a nanometer, or 10^{-15} m. As I discussed in Section I, neutrons and protons span about 1 fm, and nuclei, even the biggest ones, have radii less than 10 fm.

If you could expand a proton to the size of a pea, the atom in which it resides would be about a mile across. There is indeed a lot of space within an atom.

The smallest distances probed in high-energy experiments at accelerators are about a thousandth of a fermi (10^{-18} m). That could be said to be the smallest distance about which we know anything for sure. That limit of sure knowledge doesn't prevent theorists from imagining much shorter distances. At the so-called Planck length, which is about 10^{-35} m, or some 100 billion billion times smaller than a single proton, not just particles but space and time themselves are expected to join in the quantum dance of waves and probability (see Question 93). *Quantum foam* is the name given to this roiling broth of spacetime by the American physicist John Wheeler.* And it is at this dimension that *strings*, if they exist, writhe about to produce what we see as particles.

16. How far can one particle "reach out" to influence another one? Different forces have different ranges. Let me start with the weakest of all forces, gravity. It pulls each of us to Earth but reaches out far beyond the human domain, keeping the Moon in orbit around Earth, keeping Earth in orbit about the Sun, and holding galaxies together as giant rotating wheels. Gravity reaches from one end of the universe to the other. From a quantum perspective it does so because the force carrier, or "exchange particle" of gravity, the graviton, has zero mass. This enables it to reach out without limit, although with diminishing strength as the distance increases.

Electromagnetism is also mediated by a massless particle, the photon. And, like gravity, it reaches out without limit, also decreasing in strength as the inverse square of the distance. An important difference in the universe at large is that astronomical bodies have no appreciable net electric charge, so they don't attract or repel one another electrically, even though they continue to attract one another gravitationally. Although electromagnetism is intrinsically a much stronger force than gravity, its effects in the large are wiped out by the cancellation of positive and negative charge. Within an atom, on the other hand, electromagnetism far outcompetes gravity—so much so that gravity can be entirely ignored in the atomic realm. And how does the electrical attraction between a proton and an electron manifest itself within an atom? By the emission and absorption of photons—literally by their creation and annihilation, trillions of

---

* Wheeler loved to come up with memorable phrases. He is responsible also for the terms *Planck length* and *black hole*.

Sheldon Glashow (b. 1932), shown here lecturing on the universe at large in the early 1980s. The 2×10^{10} LY on the board refers to the then-estimated 20-billion-light-year radius of the universe. Glashow and his fellow Nobelist Steven Weinberg, both sons of Jewish immigrants, were members of the same class of 1950 at New York's Bronx High School of Science and the same class of 1954 at Cornell University. It is to Glashow that we owe the quirky name of the fourth quark, *charm*, a name that reappears in his essay collection *The Charm of Physics*. Photo courtesy of AIP Emilio Segrè Visual Archives, *Physics Today* Collection.

events every second—making the electromagnetic force, like every other force, an exchange force.

There is one other force, besides gravity and electromagnetism, that involves the exchange of massless particles. Those particles are gluons, and that force is the strong, or nuclear, force. Oddly enough, however, the strong force doesn't reach out to great distance like gravity or electromagnetism. It is confined to a distance of about 1 fm. It is not by chance that this distance is the same as the size of a proton. A proton (or neutron) has that size because gluons won't let quarks get much farther apart than that. The strong force, as it happens, gets stronger, not weaker, with increasing distance. It keeps a very tight rein on quarks, so if the three quarks within a proton started to drift farther apart, they would be pulled back with an even more powerful attractive force.*

---

*There is a subtlety here. If enough energy is poured into a proton—say in a collision in an accelerator—the quarks *can* be separated, but only if some of the

Abdus Salam (1926–1996), shown here at a physics conference in Rochester, New York, at age 29. After earning a doctorate in England, Salam returned to his native Pakistan for three years as a professor of mathematics. From his later base at Imperial College in London, where I remember the warmth of his personality and the chalk-stained gown he wore while lecturing, he reached out to promote physics throughout the developing world. Salam is the only Pakistani to have received a Nobel Prize. Photo courtesy of AIP Emilio Segrè Visual Archives, Marshak Collection.

The fourth and last in the pantheon of forces is the weak force—weak, as it turns out, relative to the electromagnetic and strong forces, but still much stronger than gravity. Its exchange particles, the Ws and Z, do have mass, a *lot* of mass. They weigh in at more than eighty times the mass of a proton. The weak force, which is responsible for the radioactive decay of many nuclei (and for the decay of a lone neutron) has a very short range, just a fraction of a fermi.

Physicists are always looking for unifications that will simplify the description of nature. One very satisfying unification—achieved by the theorists Abdus Salam, Steven Weinberg, and Sheldon Glashow in the 1960s—was between the weak and electromagnetic forces. In their "electroweak" theory, the crucial difference between the weak and electromagnetic interactions* is in the masses of the exchange particles. The massless photon as an exchange particle reaches out to human-sized dimensions or

---

energy goes into making new quark-antiquark pairs, so that new composite particles are formed and fly apart. Individual quarks are not released.

* I will be using *forces* and *interactions* more or less synonymously.

Steven Weinberg (b. 1933). Weinberg's stature among physicists is reflected in the popular but unfortunately untrue story that when he moved from Harvard to the University of Texas, he asked, outrageously, to be paid as much as the football coach. Weinberg is not only a deep theorist but also a gifted writer, as evidenced in *The First Three Minutes* and other books. Photo courtesy of AIP Emilio Segrè Visual Archives, *Physics Today* Collection.

beyond (think of cloud-to-ground lightning). The massive W and Z exchange particles* reach out to a distance less than the size of a single proton. In Section IX, where I introduce Feynman diagrams, you will see the similarity of these two interactions—much alike at the level of particle creation, annihilation, and exchange. Yet because they do differ in significant ways, they are still often described as distinct forces. The weak interaction is not only much weaker and of far less range than the electromagnetic interaction, it is also more universal. Neutral particles that experience no electromagnetic force still experience a weak force.

17. How fast do particles move? There is one particle that has no choice. The photon, being massless, has to move as fast as it is possible to move: at the speed of light (for which we use the symbol c). Other parti-

---

* These particles were discovered only after their existence was predicted by Salam, Weinberg, and Glashow.

cles can, in principle, be slowed to a crawl or brought to rest, but in practice they are more likely to be seen scooting around at not much less than the photon's speed. Cosmic-ray particles arrive from outer space at very close to the speed of light. Electrons ejected from nuclei in radioactive beta decay are nearly as fast. In the Stanford Linear Accelerator, electrons are pushed to within 0.02 meters per second (m/s) of the 299,792,458 m/s speed of light. Within atoms, electrons move at from 1 percent to more than 10 percent of the speed of light.

Bulkier entities do move more slowly. At ordinary temperature, molecules of air laze along at an average speed of around 500 m/s. We humans normally move much more slowly than that, although fighter pilots can slice through air at about that speed. Astronauts in orbit move at about 7,000 m/s, some forty thousand times slower than light.

The mind-stretching effects of relativity theory become evident only for speeds near the speed of light. In everyday life, we and the objects we deal with move much more slowly than light, so we are not normally aware of relativistic effects, just as we are not normally aware of quantum effects. Yet the speeds we regularly encounter don't differ as much from the speed of light as, say, our everyday distances and times differ from those in the subatomic domain. You probably learned in school that it takes eight minutes for light to reach us from the Sun. So, despite the apparent instantaneous illumination that results when you turn on a light, you know that light can take a while to get from one place to another in the astronomical realm. From the nearest star other than the Sun, it takes four years to reach us.

On Earth, light (in an optical fiber, for example) can get from any place to any other place in less than a tenth of a second—not a duration you are likely to notice in surfing the Web. But in the summer of 1969, we Earthlings did directly sense the speed of light (or, what is the same thing, the speed of radio waves). Those of us who listened in on the conversation between NASA mission controllers on Earth and astronauts on the Moon noticed an appreciable time lag between a controller's question and an astronaut's answer. Beyond the normal human reaction time of less than a second, there was an extra two-and-a-half-second delay as the radio waves made their way to the Moon and back.

Is the speed of light *really* nature's speed limit? There is strong reason to believe that it is, but nothing in science is absolute. Physicists have dared to speculate about particles called *tachyons* that move faster than light (and can *only* move faster than light). They have looked for them, and found none. So far the speed of light is a barrier uncrossed.

18. What are some quantum scales of time? In our everyday world, we think of the blink of an eye (perhaps a tenth of a second) as a very short time. In the subatomic domain it is an eternity. In a tenth of a second an electron can circle around the nucleus of its atom 10 million billion (10^{16}) times. For particle interactions, a useful scale of time is the time it takes light to travel across a proton, a distance of about 1 fm. This time—you can think of it as one tick of the particle's clock—is about three trillionths of a trillionth of a second (3×10^{-24} second). It is about how long a gluon lasts as it is exchanged between two quarks. Needless to say, that is not the *shortest* time that physicists have conceived of. Just as there is a "Planck length," there is also a "Planck time." This is the time it takes light to travel one Planck length. It is an unimaginably short time, about 10^{-43} second. It is the time scale where gravity and the quantum intermingle and quantum foam takes over.

Not all times associated with particles are as short as the times mentioned above. Most particles are unstable (radioactive) and some of them live long enough to move an appreciable distance in the laboratory, or even to make it from high in the atmosphere to Earth's surface. The unstable particle with the longest mean life* is the neutron, which lives, on average, fifteen minutes.† Even the muon's mean life of two microseconds (two millionths of a second) is much longer than that of most particles. Lifetimes around 10^{-10} second are common, but even so short a lifetime allows a particle to move a few centimeters and to leave a track in a detector before it expires.

---

* At Question 27 I discuss the meaning of mean life and its relation to half-life.
† Stabilized within a nucleus, a neutron can live forever.

19. What is the meaning of $E=mc^2$? In 1905, the same year in which he explained Brownian motion, introduced the theory of relativity, and proposed the photon, Einstein wrote down what has surely become the world's most famous equation, $E=mc^2$. It states that energy (E) is equivalent to mass (m)—thus the equal sign. This means that mass can be converted to energy (as it is in nuclear fission and fusion) and energy can be converted to mass (as it is when particles collide at high energy in an accelerator). It means even more: that mass and energy are really the same thing. Mass is just congealed energy, and energy has inertia (the defining feature of mass). Energy *is* mass. Mass *is* energy.

But how did the square of the speed of light (c^2) get into the act? A chunk of matter need not be moving at all, much less moving at the speed of light, in order to have mass and energy. To explain the appearance of c^2 in Einstein's equation, let me take a brief detour to a fabric store. If you buy three yards of a material priced at five dollars per yard, you know that your cost will be fifteen dollars. In equation form, $C=NP$: Your cost C is equal to the number of yards N times the price per yard P. The cost is *proportional to* the number of yards. If you change your mind and decide to buy twice as many yards, your cost will be twice as much. What will not change (for this particular material) is P, its price per yard. P is called a *proportionality constant*. It converts the number of yards N to a cost C. The "essence" of the equation is the proportionality of C to N, written $C \sim N$. The constant P does the job of converting a number of yards to a number of dollars.

In a similar way, c^2 in Einstein's equation is a proportionality constant. The essence of the equation is the proportionality of E to m: $E \sim m$. The constant c^2 does the job of converting a number of kilograms (a mass unit) to a number of joules (an energy unit). And c^2 is, in everyday units, a *very* large number, 9×10^{16} joules per kilogram (kg). A 1-kg rock thrown at a speed of $10 \, \text{m/s}$ has a kinetic energy of 50 joules, enough to do a lot of damage if it hits you in the head. Yet locked within that rock is an energy—a mass energy—of 9×10^{16} joules, more than a million billion times greater than its kinetic energy. Let's put that in perspective: The energy locked in 1 kg of mass is 1,500 times greater than the energy

This isn't exactly the way it happened. Courtesy of ScienceCartoonsPlus.com.

released in the Hiroshima atomic bomb. Less than 1 gram of mass was converted to energy in that explosion.

Does this mean that mass-to-energy conversion takes place only in the nuclear domain? No, it occurs also when you light a match or put another log on the fire or rev up your automobile engine. In those cases, however, the loss of mass is so totally minuscule that it can't be measured.

Even now, the conservation of mass (no gains or losses) is a solid principle of chemistry. Yet, at the deepest level, it is incorrect! Every time energy is released (in an "exothermic" reaction), there is some decrease in mass. Every time energy is gained (in an "endothermic" reaction), there is some increase in mass.

Even as you grasp the idea that c^2 is a constant of proportionality linking energy and mass, you may still rightly wonder what the speed of light has to do with the mass-energy equivalence. The answer starts from the fact that in relativity theory, the speed of light links space and time. (Note that speed is distance per unit of time.) Einstein brought the formerly quite disparate idea of space (measured, say, in meters) and time (measured, say, in seconds) into the single concept of four-dimensional spacetime. Had ancient scientists appreciated this unification, perhaps they would have adopted the second as a unit of distance (about 300 million meters) or the meter as a unit of time (about three nanoseconds). But they didn't, and we are stuck with space and time measured in different units. To join them into a single four-dimensional whole, we need to multiply time by the speed of light. Then ct, measured in meters, can join distance x as a full partner. As it has turned out, the other unifications in relativity theory—mass with energy, momentum with energy, and magnetic field with electric field—also need constants of proportionality, all of which include the speed of light. As to why, for mass and energy, the constant is c^2, not c, a rough-and-ready answer is that it takes the combination $(meter)^2/(second)^2$ to convert kilograms to joules.

In our everyday world, the joule is a convenient unit. In the quantum world, a more common unit is the much smaller electron volt, or eV, defined in the footnote on page 22. 1 eV is equal to 1.6×10^{-19} joule. In a cathode-ray TV tube, electrons strike the screen with an energy of about 1,500 eV (1.5 keV). Electrons in the Stanford Linear Accelerator reach an energy of 50 GeV. Protons in the Fermilab Tevatron reach 1 TeV. The current record-holder accelerator is CERN's (Centre Européen pour la Recherche Nucléaire) Large Hadron Collider in Geneva, Switzerland, which, as of this writing, has pushed protons to an energy of 3.5 TeV and is scheduled to push them to 7 TeV (a combined energy of 14 TeV when they collide).* (Even 14 TeV is still very much smaller than 1 joule.)

In the particle world, it is common to dispense with c^2 and measure masses in energy units. Thus the mass of an electron is 0.511 MeV; of a proton, 938 MeV; and of a top quark, about 172 GeV (more than three hundred thousand times the mass of an electron). There is good reason for this practice. Mass-to-energy and energy-to-mass conversions are commonplace for particles. Mass and energy are constantly being stirred together. When an unstable particle decays, for example, some of its mass energy goes into the masses of its decay products and the rest into their kinetic energy as they fly apart. How the initial energy is divided varies greatly from one particle to another. When a neutron decays, 99.9 percent of its mass reappears in the masses of the product particles, whereas when a muon decays, only 0.5 percent of its mass shows up in the masses of the products. Also, when particles crash together in accelerators, there are likely to be big conversions between mass and energy. Often a great deal of the energy poured into the particle being accelerated goes into creating new mass.

For one kind of particle transformation, 100 percent of the mass is converted to energy. When a positron (an antielectron) meets an electron, both can vanish as their mass energy is totally transformed to the kinetic energy of massless photons. In the fictional Starship Enterprise, annihilating antimatter provided the energy to drive the ship. Unfortunately, there is no realistic prospect that this source of motive power will ever be realized.

20. What is electric charge? If you were asked to describe a particular person, you could name characteristics almost without end. If you had a literary flair, you could write a book just describing one person. If, instead, you set out to write a book describing an electron, you would run out of things to say after a page or two. An electron has mass (and the energy

---

*For a table of abbreviations for large and small multipliers, see Table B.1 in Appendix B.

that goes with that mass) and it has spin. Those are its mechanical properties. In addition, it carries two kinds of charge: electric charge, which determines its strength of electromagnetic interaction, and another kind of charge that determines the strength of its weak interaction.* (There is one more kind of charge, called *color charge*, which is possessed by strongly interacting particles but not by electrons.) Here I focus on electric charge, which, more often than not, is simply called *charge*.

Charge has two extraordinarily important features. It is *quantized* (it comes in lumps) and it is *conserved* (its total amount never changes). The unit of charge is e, the amount of charge carried by a proton. We commonly measure charge in the large-scale world in coulombs, named after the late-eighteenth-century scientist Charles Augustin Coulomb, who measured the law of attraction and repulsion between charged objects. The quantum unit of charge is, to put it mildly, very small relative to the coulomb: $e = 1.6 \times 10^{-19}$ C. (To turn it around, 1 coulomb is the charge of 6 billion billion electrons.) In the quantum world, we usually take e as the unit. Using that unit, the charge of an electron is −1, of an alpha particle, +2, and, of a uranium nucleus, +92. As I have mentioned, the unseen quarks possess fractional charges in this unit: $-\frac{1}{3}$ and $+\frac{2}{3}$ (with antiquarks having charges of $+\frac{1}{3}$ and $-\frac{2}{3}$).

The conservation of charge, a rule suggested first by Benjamin Franklin well before Coulomb's measurements, means that in any process of change, the total amount of charge doesn't change; it is the same after as before. This is, so far as we know, an absolute conservation law, valid under all circumstances. I discuss some of its implications at Questions 53 and 57.

Charge is a scalar quantity—that is, it has magnitude but no direction. This means that the charge of a composite entity is simply the numerical sum of the charges of the constituents. Thus, the charge of a hydrogen atom is $1 - 1 = 0$. Stripping two electrons from a uranium atom leaves ninety-two protons and ninety electrons, for a net charge of $+92 - 90 = +2$. When a quark with charge $+\frac{2}{3}$ combines with an antiquark

---

*These "charges" are also known as *coupling constants*, for they determine how strongly the particle is coupled to other particles.

of charge $+\frac{1}{3}$, the result can be a pion of charge +1. To give one example of charge conservation in action: When an electron of charge −1 and its antiparticle, a positron, of charge +1, meet and annihilate to create a pair of uncharged photons, the net charge is zero before and after the event.

But what *is* charge? You might say that it is that certain something, that *je ne sais quoi*, that causes particles that have it to attract or repel one another, leading, in some cases, to a happy union, as when an electron and a proton join to form a hydrogen atom, and leading, in other cases, to mere deviation from a straight path, as when a rushing alpha particle passes close to an atomic nucleus. Most important, charge is what enables particles to participate in the pyrotechnic events of emitting and absorbing photons. That is the essence of what charge is. It "couples" charged particles to photons. It links matter and radiation.

21. What is spin? Nearly everyone has seen a spinning top and a rotating merry-go-round. Nearly everyone knows, too, that Earth spins around its axis once a day and revolves around the Sun once a year. In fact, there isn't much in Nature that doesn't turn. Ice skaters twirl. The Moon spins around its axis as it circles Earth (both motions completed once a month). Galaxies rotate. So do clusters of galaxies. And so do most of the fundamental particles.

The "amount" of rotation is measured by a quantity called *angular momentum*. Angular momentum is determined by how much mass is rotating, how fast it is turning, and how far the mass is from the axis of

FIGURE 9
With the right fingers showing the direction of rotation, the right thumb shows the direction assigned to the angular momentum.

rotation. It is called a *vector quantity* because it has not only a magnitude but also a direction—which is the direction of the axis, determined by a "right-hand rule" (see Figure 9).

Physicists distinguish between orbiting and spinning (or between orbital angular momentum and spin angular momentum). Earth provides an example of each kind through its daily spin and its yearly orbit (Figure 10). Particles, too, can both spin and orbit. The spin of an electron—or of any particle—is an intrinsic property of the particle. Nothing can be done to increase or decrease it. It is always there. One charged particle circling around another one, however, can have different amounts of orbital angular momentum, from zero to quite large values (large by atomic standards). An electron in a hydrogen atom, for example, has zero orbital angular momentum in its lowest-energy state of motion, and can have increasingly larger values as it moves in more highly excited states of motion.

Angular momentum, although conceptually quite different from charge, does share charge's two most important features. It is quantized and it is conserved. Historically, it also shares a quirk with charge: Its smallest value has turned out to be smaller than was originally supposed. For charge, that dicing of the quantum came with the discovery of quarks. For angular momentum, it came with the discovery of spin. In his landmark work on the hydrogen atom in 1913, Niels Bohr had postulated that all angular momenta are integral multiples of a smallest value, equal to Planck's constant divided by 2π, or $h/2\pi$. As I discussed at Question 10, that quantity is abbreviated $\hbar$ and pronounced h bar. Then, a dozen years later, along came two young Dutch physicists, Samuel Goudsmit and George Uhlenbeck, who showed how to clear up some puzzles

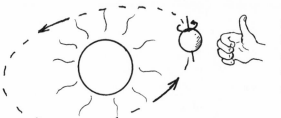

FIGURE 10
Earth has both orbital and spin angular momentum.

in atomic spectra (the pattern of light emitted by atoms). They postulated that the electron, in addition to whatever orbital angular momentum it might possess, also possesses an intrinsic spin whose magnitude is ($\frac{1}{2}$) $\hbar$. We know now that orbital angular momentum is indeed always an integral multiple of $\hbar$ (or zero), as Bohr had suggested, and that the smallest value of spin angular momentum is ($\frac{1}{2}$) $\hbar$, as Goudsmit and Uhlenbeck proposed. It is common now to use $\hbar$ as the unit of spin angular momentum, in which case we can say electrons and quarks have spin $\frac{1}{2}$, photons and W and Z particles have spin 1, and the hypothetical graviton has spin 2. There are also particles, such as pions and the hypothesized Higgs, that have spin zero.

The conservation of angular momentum means, as you might surmise, that in every process of change, the total angular momentum is the same after the process as before. To see how this works in practice, you have to know how different angular momenta add together to make a total angular momentum. Simply adding up magnitudes works for charge (a scalar quantity) but not for angular momentum (a vector quantity). Moreover, it is trickier in the quantum world because the sum of two or more angular momenta has to have a magnitude consistent with quantum rules. In the classical world, one can add a vector of magnitude 1 pointing east to a vector of magnitude 1 pointing north to get a vector of magnitude 1.41 pointing northeast. Indeed, those two vectors, depending how they point, can add up to any sum whatsoever between zero and 2—an infinite number of possibilities. Not so in the quantum world. Two angular momenta, each of magnitude 1 (in units of $\hbar$), can add to only three possible sums of magnitudes 0, 1, and 2. An electron with 1 unit of orbital angular momentum and with its intrinsic spin of $\frac{1}{2}$ can have only two possible values of total angular momentum—$\frac{1}{2}$ or $\frac{3}{2}$. More examples of granularity.

Here is one example of angular momentum conservation in a process of change: Consider an electron and a positron with no orbital motion about one another, so that their orbital angular momentum is zero, and with oppositely directed spins, so that their total spin angular momentum is also zero. Then, when they annihilate to produce two photons, the total angular momentum of the photons must also be zero, the same as before the annihilation event. Since each of the photons has spin 1, we can conclude that these two spins must be oppositely directed. Only then can spin 1 plus spin 1 equal zero.

There is one other subtle rule of combining angular momenta, which, for the sake of completeness, I need to mention. It has three parts: (1) When integral angular momenta (0, 1, 2, 3, and so forth) combine, the result is also an integral angular momentum. Thus 2 and 3 could give, among other possibilities, 1 or 4 or 5. (2) When a half-odd-integral angular momentum ($\frac{1}{2}$, $\frac{3}{2}$, $\frac{5}{2}$, and so forth) combines with an integral angular momentum, the result is a half-odd-integral angular momentum. For example, $\frac{1}{2}$ combining with 2 could result in $\frac{3}{2}$ or $\frac{5}{2}$. (3) And, finally, when two half-odd-integral angular momenta combine, the result is an integral angular momentum. An example is the electron-positron combination, whose two spins, each $\frac{1}{2}$, can result in a total spin angular momentum of zero or 1.

For common magnitudes and units of measurement in the subatomic world, see Table B.2 in Appendix B.

section IV

Quantum Lumps and Quantum Jumps

22. What are some things that are lumpy (and some that are not)?
Any intrinsic property of a particle is, almost by definition, lumpy—that
is, quantized. Such properties include the particle's mass, its charge, and
its spin—and other things that I haven't discussed yet (or have only mentioned), such as lepton number, baryon number, and color charge. Anything that characterizes the particle and can't be changed without changing entirely the nature of the particle is quantized. Properties associated
with the unconfined motion of the particle, on the other hand, are smooth,
not quantized. A particle moving freely can have any speed or momentum
or kinetic energy. Think of an automobile. Its four tires are "lumpy," like a
particle's quantized properties. The car can't have three-and-a-third or four-
and-a-half tires. Well, you might say, I could go to work on it and trim it
down to a three-wheeler or bulk it up to a six-wheeler. True enough, but the
result would bear almost no resemblance to the original car. It would be like
replacing an electron with a muon. You would have a new entity. Moreover,
the number of tires, even after the overhaul, would be integral, not fractional. And, whether overhauled or not, the vehicle could move down the
highway at any speed whatsoever and travel any distance whatsoever.
Speed and distance would be its "smooth" (nonquantum) attributes.

When motion is confined—as it is, say, for an electron in an atom or
a proton in a nucleus—certain properties related to that motion are
also quantized. An electron in an atom has not only a particular mass,

50

charge, and spin, it also has an energy of motion and an orbital angular momentum that are lumpy. At Question 66 I discuss how the wave nature of the electron can explain the lumpiness associated with confinement. It's as if your car, although capable of any speed on the open road, was confined to only certain speeds when racing on an oval track.

The general rule is that any property that is confined (either to a single particle or to a small region of space to which the particle's motion is restricted) is lumpy, or quantized; and any property that is unconfined (such as the speed of a freely moving particle) is smooth, or unquantized. Now enters a subtlety. No particle is *truly* unconfined. There are always *some* limits to its motion. Does this means that *all* properties of particles are quantized? In principle, yes. In practice, no.

Consider an oxygen molecule in the air in your living room, confined by the room's walls. Overlooking the molecule's frequent collisions with other molecules, a physicist can calculate its allowed (quantized) energies (see Figure 11). These prove to be so closely spaced that it is impossible, in practice, to tell the difference between one allowed energy and the next. Jumping sequentially through a set of neighboring energy values is indistinguishable, in practice, from a smooth variation of energy. So, for practical purposes, it makes no difference that the energy values are quantized. Classical physics works just fine to describe the molecule's motion.

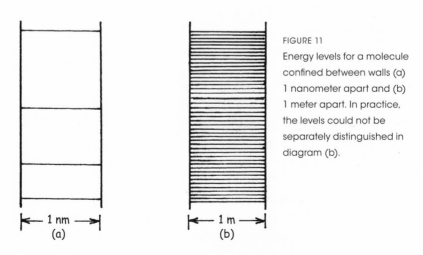

FIGURE 11

Energy levels for a molecule confined between walls (a) 1 nanometer apart and (b) 1 meter apart. In practice, the levels could not be separately distinguished in diagram (b).

Here is one more example from the large-scale world: Suppose you are standing on the sideline at Junior's soccer game, and you turn to watch Junior run down the field. Because you are turning, you have angular momentum, which has to be an integral multiple of the smallest value $\hbar$. But what an integer—perhaps a billion trillion trillion (10^{33})! Being a student of physics, you reason that if you turn more slowly—much more slowly—you might be able to reduce your angular momentum so much that its quantum nature becomes evident. It's hopeless. You are stuck in your classical world. If you turn only your head, not the rest of your body, and take a full hour to turn it from left to right, your angular momentum will still be trillions of trillions of times larger than $\hbar$. Increasing or decreasing your angular momentum by 1 quantum unit would be impossible to do and impossible to detect.

Despite these examples, there are instances in which quantum effects are felt in the large-scale world. I mentioned one, superconductivity, at Question 2. Another, with enormous practical implications, is the behavior of electrons in metals. In a metal there are vast numbers of electrons, liberated from bondage to individual atoms, that are free to roam within the confines of the material. Each electron, like the oxygen molecule in the living room discussed above, is restricted to certain very closely spaced energy values. There are an enormous number of possible energy values, but there are also an enormous number of electrons. Because of the exclusion principle, which I discuss at Question 25, only two electrons (corresponding to the two possible directions of its spin) can have any one of these energy values. So the electrons "pile up," occupying ever higher energy states. Although the spacing between adjacent allowed energies is a minuscule fraction of an electron volt, electrons at the top of the "pile" have energies of several electron volts. It is these energetic electrons—held aloft, so to speak, by the exclusion principle—that account for the metal's ability to readily conduct electricity.

23. What is a "state of motion"? In the classical world, we can specify the motion of an object by saying where it is, how fast it is moving, and in what direction. You could get out your cell phone, call a friend, and say,

"Here is my state of motion. I am on the northwest corner of 45th and Madison, strolling north on Madison at two miles per hour." Your friend, if so inclined, could calculate your momentum and kinetic energy. No such specificity is allowed in the quantum world. A particle's location can't be exactly specified, nor can its speed or direction of motion. But all is not fuzzy. Certain "global" properties of the particle's motion can be pinned down exactly, such as its energy and its angular momentum and perhaps the direction of its spin axis. These global properties define what we call its *state of motion*. Going back to the large scale, it's as if you call your friend from your car to report, "Here is my state of motion. I am circling a two-block area on the Upper East Side at an average speed of ten miles per hour looking for a parking spot."

The concept of state of motion we owe to Niels Bohr, who, in his famous 1913 work on the hydrogen atom, introduced the idea of a "stationary state." By *stationary* he did not mean motionless. He meant that certain properties of the motion, such as energy and angular momentum, were fixed—fixed as long as the electron was in that state of motion, which might be for a short or a long time. But in Bohr's telling, the electron did not ooze from one state of motion to another. Its properties stayed fixed until the electron jumped suddenly to another state.

Bohr also recognized that in an atom (and, as we now know, within a nucleus or any other confined system) there is a state of lowest energy. We call that the *ground state*. Although a particle in that state has energy, it is useless energy. It is also called *zero-point energy*, the energy that a system still possesses after all possible energy has been drained from it. An object brought to an absolute zero of temperature would have zero-point energy and nothing else. Because of zero-point energy, there is indeed such a thing as perpetual motion.

Bohr recognized that when an atom is not in its ground state, it is in some more energetic "excited state." An excited state is simply any state of motion with more energy than the ground state. In a given atom there are actually an infinite number of excited states. The first few such states are well separated in energy from the ground state—their discreteness is obvious. This means that their quantum nature is obvious. As the energy increases, the states cluster ever closer and closer together, until finally they look like a "continuum"—that is, they resemble what would be expected classically (see Question 3).

The wave nature of matter helps clarify why stationary states of motion exist. Here is a very quick description of the linkage between waves and stationary states. (I expand on this idea at Question 67.) We know that the strings of a piano or a harp or the column of air in a flute or a bugle vibrate at only certain frequencies, emitting particular tones of sound. To be specific, consider a bugle. The bugler can produce various tones by blowing in different ways, setting up different wave patterns within the bugle. Each tone is, in effect, a musical "stationary state." The bugler can't make any tone whatsoever. He is limited to the tones permitted by these particular vibrations. In a somewhat similar way, the wave associated with an electron in an atom can vibrate only in certain ways, and these allowed vibrations provide the stationary states.

Actually, there is a way in which the bugle is strikingly analogous to the hydrogen atom. There is a lowest tone that the bugle can emit—its "ground state," so to speak. Its second tone is well separated from its first, its third tone a little less well separated from its second. As the bugler makes sounds that are more and more high-pitched, they get closer and closer together on the musical scale, just as more highly excited stationary states in a hydrogen atom get closer and closer together in energy.

24. Is a hydrogen atom in an excited state of motion the same atom in a different state or is it a different atom? This is a subtle and deep question. Now and then in the book I ask you to fasten your seatbelt. Here is one such place.

You are likely to say, "Well, surely an atom in an excited state is still the same atom. Its energy has changed and, as a result, its mass has changed a tiny bit, but it's still the same electron moving around the same proton." To argue otherwise would seem like painting your house and then calling it a different house. Most practicing physicists and chemists would in fact agree that it makes sense to treat the excited atom as a different state of the same atom. Everything works out all right if you look at it in that way. And it's convenient to do so. Yet, at the deepest level, *every entity that differs in any particular from another entity is a distinct entity.* Quantum physics does not distinguish in principle between (1) an atom that emits a photon as it jumps from a higher to a lower energy state, and (2) a pion that emits a neutrino as it transforms itself into a muon. For the latter process, all would agree that the pion, the muon, and the neutrino are distinct particles, and that when one of them disappears (is annihilated), the other two appear (are created). The logic of quantum physics requires that, however convenient and "obvious" it might be that the hydrogen atom exists in two states, one before and one after the photon is emitted, in fact the atom in its excited state, the atom in its ground state, and the photon are three distinct entities, and when the excited atom disappears (is annihilated) the ground-state atom and the photon appear (are created).

Taken strictly, quantum physics does not even allow us to say that a hydrogen atom is "composed of" a proton and an electron, any more than it allows us to say that a pion is composed of a muon and a neutrino. If, hypothetically, you drop a proton and an electron into an empty box, after a while the box might contain a hydrogen atom. That atom will not have a mass equal to the sum of the masses of the original particles. Its mass will be slightly less. And there is no way in principle to measure the mass of an individual particle—proton or electron—within a hydrogen atom. The atom is a new entity. You might understandably consider this to be splitting hairs or even to be a deranged view of the world, like saying that a house into which someone walks is not a house plus a person, but has become something else entirely. The quantum physicist has to say, "Yes, it has." He or she would ask you to imagine dropping a muon and a neutrino into that hypothetical empty box. After a time, the box might contain a pion. We are not tempted to say that the pion consists of a muon and a neutrino, for its mass is quite different from the sum of the masses of those two particles, and it has other properties that differ dramatically as well. Saying a house into which a person has walked is not a house plus a person but something entirely new defies common sense, but it is not the only aspect of quantum physics that leaves common sense behind.

Having said all of the above, I now have to tell you that in fact the physicist *does* say that a hydrogen atom consists of a proton plus an electron, and *does* say that an excited atom and a ground-state atom are two states of the same entity. It all has to do with the percentage change between one state and another, or one entity and another. When that percentage change (of mass, say) is quite small, as it is when an electron and a proton join to form a hydrogen atom or when an atom is elevated from its ground state to an excited state, it is useful and quite adequate to think of the atom as a composite and of the excited state as a manifestation of the same atom. But when the change from one state to another or one entity to another is dramatic, it makes more sense to adopt the strict quantum interpretation that one is dealing with different entities.

25. What are quantum numbers? What are the rules for combining them? Anything that is lumpy can be numbered. If your car were capable of driving only at exactly ten miles per hour, or at twenty, or at thirty, and so on, but nothing in between those speeds, you could just number your speeds: Speed 1, Speed 2, Speed 3, and so on. If—as is indeed the case—the orbital angular momentum of an electron in an atom can have only the values 0, $\hbar$, $2\hbar$, $3\hbar$, $4\hbar$, and so on, then those values can be labeled by the integers 0, 1, 2, 3, 4, and so on. Those are called *quantum numbers* because they number a quantized property. In fact, it is common to say that the orbital angular momentum *is* 2 or 3 or whatever integer, meaning it is that integer multiplied by $\hbar$.

The standard notation for the orbital angular momentum quantum number is ℓ (script "ell"), and the rule for it is that it is equal to zero or a positive integer:

$$\ell = 0, 1, 2, 3, \ldots.$$

It turns out that for the ground state of the hydrogen atom, $\ell = 0$. For the first excited state, ℓ can be 0 or 1. For the second excited state it can be 0 or 1 or 2. There is no limit to how large ℓ can be, and when it gets to be quite large, the correspondence principle kicks in as the quantum behavior begins to resemble classical behavior. A way to help see this is to notice that going from $\ell = 1$ to $\ell = 2$ is a 100 percent change, whereas going from $\ell = 100$ to $\ell = 101$ is a one percent change, and small fractional changes lead to near-classical behavior.

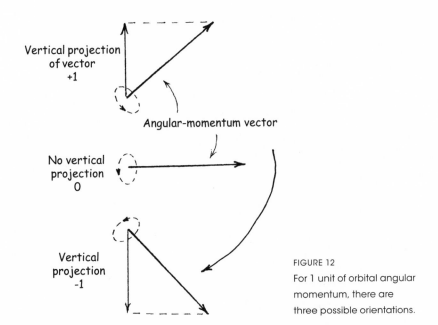

Vertical projection
of vector
+1

Angular-momentum vector

No vertical
projection
0

Vertical
projection
-1

FIGURE 12

For 1 unit of orbital angular
momentum, there are
three possible orientations.

Now to a truly remarkable feature of angular momentum. Not only is its magnitude quantized, so is its orientation. Specifically, the component of angular momentum in a given direction can have only certain values, and these values differ, one from the next, by exactly $\hbar$. The way this works is shown in Figure 12. If an electron has an orbital angular momentum of 1 quantum unit ($\hbar$), the component of that angular momentum can be $\hbar$, 0, or $-\hbar$—three values separated by $\hbar$. Roughly stated, the electron's angular momentum in this case can point up, down, or sidewise. As shown in Figure 13, if the angular momentum is $2\hbar$ there are five possible orientations. Since the angular momentum component is lumpy, it, too, can be assigned a quantum number. We call that quantum number m_ℓ. Its values can range from ℓ to $-\ell$, in steps of 1. Thus, for 1 unit of angular momentum, m_ℓ can be 1, 0, or -1. For 2 units of angular momentum, m_ℓ can be 2, 1, 0, -1, or -2.

But how does the physicist (or the atom) decide which direction is up? Who—or what—chooses the axis along which the components will be measured? The startling answer is "It doesn't matter." If you measure

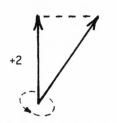

FIGURE 13

For 2 units of angular momentum, the five orientations also have components separated by 1 unit.

+2

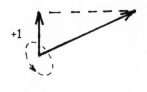

+1

the component of angular momentum along any axis, it will be found to have an integral quantum value. If you measure it along some other axis, tilted at any arbitrary angle to the first axis, it will again show itself to have an integral quantum value—not necessarily the same one. According to quantum theory, if you *don't* measure it, there is no way of saying what the component might be—it is unpredictable and unknowable. This is a consequence of one of the most mind-stretching of all the ideas in quantum physics, the idea of *superposition,* that a system can exist in two or more states of motion at the same time. Applied to angular momentum, it's as if a satellite circling Earth could be following a number of different orbits at the same time, with various tilts relative to the Equator. I pursue the idea of superposition at Question 76.

0

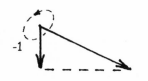

-1

For an electron in an atom, there is a third quantum number, called the *principal quantum number* and designated n. Perhaps it is called *principal* because it was the first quantum number to be identified—not sur-

-2

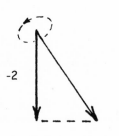

prisingly, by Bohr, in 1913. He assigned $n=1$ to the ground state of the atom, $n=2$ to the first excited state, $n=3$ to the second excited state, and so on. Now we understand the real meaning of n. It measures the total number of cycles of oscillation of the electron wave in the atom, both in-and-out (radial) and roundabout (tangential). For example, for quantum number $n = 20$ there is one state of motion for which the electron wave executes twenty cycles of oscillation in the radial direction (in and out from the nucleus) and is smooth without oscillation in the tangential (sometimes called the *azimuthal*) direction. That wave looks a bit like the ripples in a pond around a place where a pebble fell in. There is another $n = 20$ state in which there are nineteen cycles of oscillation going tangentially around the nucleus and only a single cycle in the radial direction. That wave could be simulated by a rope laid out in a horizontal circle and bunched up into little peaks and valleys nineteen times around the circle. In between those two states of motion are eighteen others with various combinations of in-and-out and roundabout wave oscillation. In all, there are twenty different states of motion with $n = 20$ (each with its own value of the angular momentum quantum number ℓ).

Wolfgang Pauli *(left)* (1900–1958), shown here with Niels Bohr and a tippe top in Lund, Sweden, in 1954. Pauli could be sharp in his criticisms, as in his famous put-down of a muddled paper, "It's not only not right, it's not even wrong." To some young theorists I knew in Göttingen, he typically wrote "total nonsense" in response to any of their new ideas—then, a week later, thawed and congratulated them. Once, when lab equipment in a Göttingen laboratory broke without apparent cause, it was attributed to the "Pauli effect": Pauli, it turned out, was on a train passing through Göttingen at the time. When Pauli, terminally ill in a Zurich hospital, was visited by an assistant, he reportedly said, "Did you notice my room number? It is 137." (See Question 94.) Photo by Erik Gustafson; courtesy of AIP Emilio Segrè Visual Archives, Margrethe Bohr Collection.

As this discussion implies, the lowest-energy state, or ground state, with $n = 1$ has just a single cycle of oscillation for its wave (actually a rise from a small value at one "edge" of the atom through a maximum value at the atom's center, diminishing to a small value again at the atom's opposite "edge"). It is a rather featureless wave "blob." I remark here, too, that to the extent that protons and neutrons move somewhat freely within a nucleus, they, too, are described by the same quantum numbers as electrons in atoms.

By 1925, a dozen years after Bohr's pathbreaking work, physicists had embraced the three quantum numbers n, ℓ, and m_l that I have discussed, and they understood that each separate state of motion of an electron in an atom is characterized by a specific combination of these three numbers (although they had not yet arrived at the wave picture of the electron state). In that year, the twenty-five-year-old Austrian physicist Wolfgang Pauli (then working in Germany) offered a double-barreled insight into quantum states and quantum numbers. First, he advanced what has come to be called the *exclusion principle*, the postulate that no more than one electron can occupy a given state of motion at any time—or, equivalently, that no two electrons can have the same set of quantum numbers at the same time. Second, he said, to make this principle compatible with all that is known about atomic spectra, a fourth quantum number is needed. This new number, he noted, has the strange property that it can take on only two values, rather like a toggle switch.

Then, almost immediately, two Dutch physicists, Samuel Goudsmit and George Uhlenbeck, whom I introduced at Question 21, offered a daring vision of what that fourth quantum number represents: spin—or, more specifically, the orientation of spin. The electron, they said, must

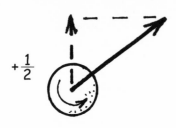

$+\frac{1}{2}$

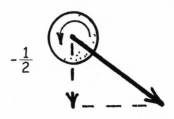

$-\frac{1}{2}$

FIGURE 14

The two possible orientations of electron spin have components that differ by 1 unit.

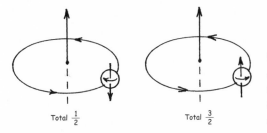

have $\frac{1}{2}$ unit of spin. Consistent with the quantum rule that successive components of angular momentum must differ by 1 unit, the electron spin has only two possible orientations, with components $+\frac{1}{2}$ and $-\frac{1}{2}$ (which, indeed, differ by 1). An electron's two possible spin directions are shown in Figure 14. The spin itself is a quantized quantity, so it can be assigned a quantum number, which we can call s. Like mass and charge, however, this is a fixed property of the electron that has only one value, so there is no need to keep track of it. The orientation quantum number (m_s), on the other hand, can take on two values. That is Pauli's fourth quantum number.

So an electron in an atom, occupying any particular state of motion, is labeled by four quantum numbers: n, ℓ, m_ℓ, and m_s. Since protons and neutrons also possess $\frac{1}{2}$ unit of spin, they, too, are characterized by the same set of quantum numbers when they move around within a nucleus.

At Question 21 I presented the rules for combining angular momenta. Figure 15 shows how one of these rules can be applied for an electron with orbital angular momentum 1 and spin angular momentum $\frac{1}{2}$. The electron's total angular momentum can be $\frac{1}{2}$ (with two possible orientations) or $\frac{3}{2}$ (with four possible orientations).

26. What is a quantum jump? Back in 1913, Niels Bohr advanced some of the most important ideas of quantum physics, ideas that are still with us and still seem startling. One of these is the idea of a quantum jump—a sudden leap from one state of motion (or stationary state, as Bohr called it) to another. What could be more divorced from common sense? One might even ask, "What could be more absurd?" It's as if you were circling the block on the Upper East Side of Manhattan, patiently looking for a parking space, only to suddenly find yourself driving around Washington Square in Greenwich Village, without having traveled from one place to the other. When Bohr's mentor Ernest Rutherford first saw this idea in a draft of Bohr's paper, he was understandably bothered by it. "It seems to me," he wrote to Bohr,* "you would have to assume that the electron knows beforehand where it is going to stop." And Rutherford could have added "and when it is going to jump." The when and the whither can't be known in advance. The quantum jump is something physicists continue to brood about nearly a century after Bohr introduced the idea. Albert Einstein repeatedly said he didn't like it, and other physicists have said, in effect, "Well, we don't really like it either, but it is a fact of the quantum world."

The thing that sets a quantum jump apart from other transitions is that it is spontaneous. Nothing causes it. Nothing triggers it. It just happens.† Despite its unpredictable abruptness, a quantum jump follows all the rules. In particular, various quantities such as energy and charge and angular momentum are conserved—that is, they are the same after the jump as before it. When an electron, for example, jumps from a higher-energy to a lower-energy state of motion in an atom, it emits a photon, which carries away the energy difference. (Bohr himself had not yet accepted Einstein's photon idea, but he recognized that the atom's loss of energy would be compensated by energy added to radiation.)

Now we recognize quantum jumps in other contexts than the one in which they were introduced. Any spontaneous transition from a higher-energy to a lower-energy state is a quantum jump, whether or not a photon is emitted. A radioactive nucleus, for example, can suddenly eject an alpha particle, transforming itself into another nucleus with two fewer protons and two fewer neutrons. The energy difference between initial and final states (really a mass difference) goes into the kinetic

---

*Bohr was in Denmark and Rutherford in England at the time. Communication was by mail, which, when there was no ocean to cross, was often swifter than it is now.

† "Stimulated emission" provides an exception to this rule. See Question 80.

George Gamow (*left*) (1904–1968), shown here with Wolfgang Pauli on a lake steamer in Switzerland in 1930. Both were in their twenties at the time and both had already made important contributions to physics. Their personalities were as different at their taste in clothes. Photo courtesy of Niels Bohr Archive, Copenhagen.

energy of the emitted alpha particle (and, to a lesser extent, into the kinetic energy of the recoiling nucleus). Decays of unstable particle are also quantum jumps, with the same unpredictability of when the event happens—and, sometimes, unpredictability of what the products of the decay will be. When a pion disappears, for instance, and a muon and a neutrino fly away from the place where the pion was, that, too, is a quantum jump.

In 1928, when George Gamow,* a Russian physicist then working in Copenhagen, and Ronald Gurney and Edward Condon† (respectively, British and American) offered a theory of radioactive alpha decay, they

---

* I got to know Gamow at Los Alamos in 1950. By that time, he was as well known for his prankster's sense of humor and his consumption of strong beverages as for his brilliance as a physicist. He added both good ideas and sparks of levity to the group.
† Condon was later director of the National Bureau of Standards in Washington, but ran afoul of Senator Joseph McCarthy and had to leave government service, first for an industrial research lab, then for a university professorship. "My problem," I once heard him say, "was that I joined every organization whose goals I approved of, without asking who else might belong to it."

described a new kind of quantum jump—barrier penetration, or tunneling. A barrier that, according to classical physics, is impenetrable can nevertheless be breached in a spontaneous quantum jump if getting through it releases energy. Alpha decay involves just such barrier penetration, and it shows up, too, in some modern devices such as the scanning tunneling microscope and the tunneling diode.

27. What is the role of probability in quantum physics? The simple answer is "a starring role." Much that defies everyday common sense in quantum physics has to do with probability. Consider an electron perched in an excited state in a hydrogen atom. As suggested in Figure 16, the electron has two things to "decide"—when to jump and, if there is more than one lower-energy state available to it, where to jump. Both "decisions" are governed by probability. The quantum physicist can calculate the chance that the electron will jump to each of the lower-energy states, and can also calculate the average time that the electron will occupy the excited state before it jumps. (For more complicated systems, such as a radioactive nucleus, we don't know how to calculate the probabilities, but they are surely present and surely govern what happens.)

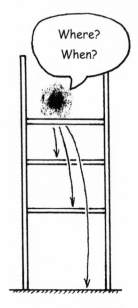

FIGURE 16

Probability governs the where and the when of a quantum jump.

Although the circumstances of the quantum jump—the when and the where—are uncertain, there is one thing that is certain: *Every* atom of the same kind in a given excited state is governed by the *same* probabilities as every other atom in that state. If you have a million atoms, all in a particular excited state, they will make quantum jumps to lower-energy states at a million different times. Yet each of them is ruled by the same inexorable probabilities. If you measure those million lifetimes and average them, you will get a "mean lifetime" for that particular state. Then if you repeat the experiment with a million other atoms and again average their lifetimes, you will get (very closely) the same mean lifetime. The mean lifetime is a built-in characteristic of that state. Possibly no individual atom will ever live for a time exactly equal to the mean lifetime, but if you measure enough quantum jumps from a particular state, you will get, reliably that same mean lifetime.*

As another example, consider a particle called a *lambda,* somewhat more massive than a neutron or proton. It is an uncharged baryon[†] with a mean life of 2.6×10^{-10} second (a bit less than a nanosecond), and it decays in two principal ways, either into a proton and a negative pion or a neutron and a neutral pion:

$$\Lambda \rightarrow p + \pi^- \qquad 64 \text{ percent of the time}$$

or

$$\Lambda \rightarrow n + \pi^0 \qquad 36 \text{ percent of the time}$$

These are the lambda's two principal quantum jumps (there are other possibilities of much lower probability). Even though the initial and final particles are very different, these are as surely quantum jumps as the ones first visualized within the hydrogen atom by Bohr. A quantum jump, strictly defined, is any spontaneous transition in which mass decreases (yes, in an atom, the mass decreases—slightly—when a photon is emitted). As in the atom, probability plays a role by dictating not only the mean lifetime but also the choice of final state. The ratio of the relative percentages mentioned (64/36) is what is called a *branching ratio.*

The decrease of mass that accompanies every instance of spontaneous decay can be called the *downhill rule.* Just as a rock rolls downhill, not uphill, quantum jumps proceed downhill in mass. But energy in total is always conserved. For the rock rolling down a hill, its loss of poten-

---

* Mean lifetime and decay probability are the inverse of each other. For example, if the chance of the radioactive decay of a particular nucleus is 0.25 per hour, the mean lifetime for that nucleus is four hours. (This does *not* mean that all of the nuclei will have decayed after four hours. It means that at the end of any hour, about ¼ of those present at the beginning of the hour will have decayed and ¾ will not have decayed. After four hours, about ⅓ of the original batch of nuclei will still be there.)
† For a definition of *baryon,* see Question 44.

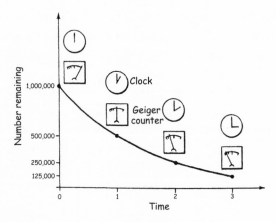

FIGURE 17
The exponential function describes both the number of remaining radioactive nuclei and the rate at which they decay.

tial energy is matched by gains in kinetic energy and heat energy. For a decaying lambda particle, the loss of mass is matched by the kinetic energies of the fragment particles.

Probability in decay processes reveals itself not only in the unpredictability and seeming randomness of individual events, but in the overall pattern followed by a large sample of unstable particles. Over time, the number of particles that have not yet decayed follows a particular mathematical law described by an *exponential function*. Such a function is shown in Figure 17. If one starts with a million radioactive nuclei, about 500,000 will remain after a time called the *half-life*. After a somewhat longer time, which is the *mean life*, about 37 percent will remain.* It turns out, interestingly, that the *rate* at which the decay events are taking place follows the same exponential law.

As far back as 1899, Ernest Rutherford, a pioneer in the study of radioactivity, observed that some of the radioactive atoms he was studying followed an exponential law of decay, and he correctly inferred that a

---

*The exact ratio of half-life to mean life is 0.693—this is true for events such as quantum jumps where the chance of decay doesn't change as time passes. For humans, whose life spans are also governed by probability, the situation is very different. The number of survivors does not follow an exponential curve. For Americans in 2004, the half-life was about eighty-one years, actually longer than the mean life of about seventy-seven years.

law of probability must be at work. He had no inkling, however, that it was a fundamental probability that he was observing. Probability was, after all, a very familiar concept in everyday life and was well known in science. That "classical" probability we can call a *probability of ignorance.* It comes about whenever not enough is known to predict an outcome with certainty. We can't predict whether a coin flipped at the beginning of a football game will land heads or tails, because we are ignorant of many factors: what force and torque were applied to it, from how high it was flipped, what air currents may have influenced it on its way down. Because we lack that information, all we can say is that there is a fifty–fifty chance of the coin landing heads or tails.

Rutherford understandably assumed that complex events were taking place within each atom, events of which he was ignorant, and it was these factors that produced the observed probability in the radioactive decay. Even after Bohr's stationary states and quantum jumps were accepted, even after quantum physics was well launched, physicists were reluctant to let go of the idea that the probability they observed was a probability of ignorance. It was an act of bravery for the German physicist Max Born to suggest in 1926 that in quantum physics a *fundamental probability* is at work. What Born proposed, and what we now accept, is that no matter how much you know about a quantum system, its behavior is still subject to laws of probability. It is a daunting idea that even if you know everything that can be known about a particular system, you cannot predict what it will do, and that if two systems are completely identical, they may behave in different ways. It's as if at a bridge tournament, identically prepared decks were provided to two tables and then the hands that were dealt were not the same.

The probability inherent in quantum physics troubled Einstein throughout his life. He liked to say that he could not believe that the good Lord played dice and suggested that only a malicious God would introduce a fundamental probability into nature.*

---

* Although Einstein frequently invoked God, or "the Old One," he stated more than once that he did not believe in a personal God.

"God is subtle but not malicious" ("Raffieniert ist der Herr Gott / Aber Boshaft ist Er nicht"). This quotation from Einstein is carved above a fireplace in Jones Hall (formerly Fine Hall) at Princeton University. Photo courtesy of Denise Applewhite.

28. Is there any certainty in the quantum world? The short answer is: Yes, lots. All of the quantum numbers I have discussed are solid. Mass, charge, spin, energy, lepton number, and baryon number can all be established with fixed values for any particle or state of motion. And, paradoxically, even probability is certain. The radioactive nucleus of a fluorine-18 atom, for instance, has a mean life of about 9,500 seconds.* In any given second, its chance of decay is 1 part in 9,500. This is true of each and every such nucleus, and is true no matter how long that nucleus has already existed. In short, the probability 1 per 9,500 per second is a fixed and certain property of that nucleus. Some quantum probabilities can be calculated. Most are known only through measurement.

---

* Flourine-18, which emits a positron, or antielectron, as it decays into oxygen-18 is widely used in positron emission tomography (PET scans).

Reliably definite probabilities are well known in the large-scale world. The chance that a balanced coin will land heads-up is exactly 50 percent. The chance that a pair of thrown dice will come up "snake eyes" (two "ones") is exactly one in thirty-six. In a similar way, in the quantum world, there is certainty in uncertainty.

section V

Atoms and Nuclei

29. What is a line spectrum? What does it reveal about atoms?
Long before there were photons, there was light. That is to say, scientists knew a great deal about light long before they recognized its quantum nature. In the seventeenth century, Isaac Newton used a prism to separate white light into its constituent colors and called the fan of colors a *spectrum*. In the century that followed, scientists used prisms to study not only sunlight and candlelight, but also light emitted by specific substances, and they found that the relative intensity of different colors depends on the substance. A flood of discoveries followed in the nineteenth century, beginning with Thomas Young's demonstration of light interference in 1801, clear evidence that light has a wave character.

Any device that spreads light apart into a spectrum of colors is called a *spectroscope* (see Figure 18). By the mid-1800s, the spectroscope had become a precision instrument. Light admitted into it through a narrow slit strikes either a prism or a diffraction grating, the effect of which is to deflect light of different wavelengths in different directions.* It is then

---

* The prism, usually made of glass, acts by slowing light of different wavelengths by different amounts. The grating, a series of very fine parallel lines ruled on glass or metal, acts by producing a different interference pattern for each wavelength of light. In both cases, the direction of the emerging light depends on its wavelength.

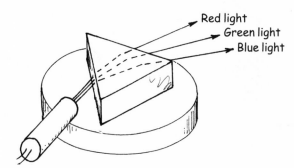

Red light
Green light
Blue light

FIGURE 18
In this spectroscope, a prism sends light of different wavelengths in different directions.

possible to measure the wavelengths. By the end of the nineteenth cen-tury, wavelengths were the most accurately measured quantities in phys-ics. Given a wavelength, and given a wave speed (the speed of light was measured with ever increasing precision in the nineteenth century), a scientist or student of physics can calculate a vibrational frequency of the wave. By the time of Planck's work in 1900, measurements had been extended into the infrared and ultraviolet regions.*

Light from an incandescent bulb is a "continuous spectrum." That means that it consists of a continuous range of frequencies spread smoothly across the spectrum. Light from the Sun is *almost* a continuous spec-trum, but not quite. It also exhibits a "line spectrum." In 1814, Joseph von Fraunhofer, a German optician, discovered a set of dark lines in the solar spectrum, fittingly known today as the *Fraunhofer lines*. A "line" in a spectrum is an image of the spectroscope's entrance slit at a particular frequency (or wavelength). For the dark lines observed by Fraunhofer, it shows the absence, or near absence, of light of that frequency. Scientists later discovered bright lines, or "emission lines," in laboratory spectra on Earth showing enhanced radiation at particular frequencies.

---

*For the record, the range of wavelengths for visible light is from about 400 nm to about 700 nm (a bit less than an octave, in musical terminology), and frequencies are in the range of about 4.5 to 7.5×10^{14} hertz (1 hertz being one cycle per second). Infrared radiation has lower frequency and longer wavelength than visible light; ultraviolet has higher frequency and shorter wavelength than visible light.

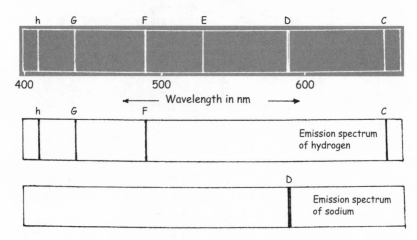

Solar Absorption Spectrum

FIGURE 19 Spectral lines emitted by sources on Earth match absorption lines in sunlight.

Some of the emission lines turned out to match precisely Fraunhofer's dark lines. This and other evidence showed that a material can emit and absorb light at the same frequencies. Moreover, every element has its own particular "signature" of spectral lines.* In the mid-nineteenth century, this made spectral analysis a tool of chemical analysis. Today it makes it possible for astronomers to identify elements in distant galaxies or gas clouds. Fraunhofer's lines provided some of the first evidence that the same elements exist in the cosmos as on Earth. These lines result from the absorption of sunlight at selected frequencies by cooler atoms hovering above the Sun's surface. Interestingly, one element was discovered through one of its spectral lines in sunlight before it was identified on Earth. Observing sunlight during an eclipse in 1868, the British astronomer Norman Lockyer and the French scientist Pierre Janssen independently noticed a new spectral line, which Lockyer cor-

---

* A given element's absorption and emission spectra are not identical, because absorption takes place "upward" in energy primarily from an atom's ground state, whereas emission takes place "downward" in energy, not only to the ground state but to other lower-energy states as well.

rectly attributed to a new element. Appropriately, he named it *helium*, after the Sun.* Figure 19 shows some emission spectra and the principal lines in the solar spectrum.

Many of the greatest advances in science have come from uniting things that had been believed to be separate. Newton united motion on Earth with motion in the heavens. Einstein united space and time as well as mass and energy. Early in the nineteenth century, in a Danish classroom, Hans Christian Oersted had revealed the link between electricity and magnetism. Later in that century, the British physicist James Clerk Maxwell added another link, uniting light and electromagnetism. When electric charge oscillates or is accelerated, Maxwell argued, electromagnetic radiation is emitted. At certain frequencies—a few times 10^{14} hertz—that radiation is in the form of visible light. At lower frequencies, it is in the form of radio waves (first demonstrated by Heinrich Hertz in 1886 and now filling up space in our wireless world). At greater frequencies it is in the form of X-rays and gamma rays. Across the spectrum the mechanism is the same. An oscillating charge can give up its energy to produce electromagnetic radiation. That radiation, in turn, can give up its energy to cause charge to oscillate.

So, according to Maxwell, light is just another electrical phenomenon. When charge oscillates in an antenna at, say, 1 million cycles per second, radio waves of frequency 1 megahertz are emitted (in the middle of the AM band). When charge oscillates within an atom at, say, 6×10^{14} hertz, green light is emitted (in the middle of the band of visible light). Or such is the classical explanation of an atom's radiation.

As the nineteenth century drew to a close, physicists—at least those who believed in atoms—had no reason to be puzzled by line spectra. Natural frequencies of oscillation were well known throughout the mechanical world, and were by then well known in the electrical world as

---

*Not until 1895 was helium discovered on Earth. Interestingly, Earth's helium is largely the result of radioactive alpha decay. An alpha particle, once slowed, captures two electrons and becomes a helium atom. Helium is found in small amounts imbedded in crystalline material and in larger amounts as a component of natural gas.

well through the phenomenon of radio transmission. Following Thomson's discovery of the electron, it was easy to visualize electrons as constituents of atoms. Scientists could reasonably assume that when an element emitted certain frequencies of light, it was because electrons in the atoms of those elements oscillated at just those characteristic frequencies, much as radio waves are emitted by an antenna in which electrons are oscillating at certain characteristic frequencies. And absorption operated as the reverse of emission. Radio waves could set electrons in an antenna into oscillation. Light waves could set electrons within atoms into oscillation. Since the electrons could be expected to oscillate only at particular "natural" frequencies, emission and absorption of light would occur only at certain frequencies, explaining line spectra.

How do we look at it now? In some ways, not so differently, since we still say that light is emitted and absorbed by electrons in atoms and that energy is exchanged between the electrons and the light. But in very important ways, our viewpoint has changed dramatically. There are two vital differences between the classical and quantum explanations of line spectra. One brave step that Bohr took in his theory of the hydrogen atom was to free himself from the classical idea that the vibration frequency of the electric charge matches the radiation frequency—a rule implied by Maxwell's theory and well established for radio waves. He reasoned that since the electron vibrates at one frequency in its initial state and a different frequency in its final state, the radiated frequency can't match the electron frequency. Instead, the radiated frequency is governed by the energy change, as implied by Planck's formula $E = hf$. Just rearranging it slightly, we get

$$f = \Delta E / h,$$

which says that the radiated frequency f is equal to the energy change ΔE divided by Planck's constant h. The electron vibrational frequency doesn't enter into it.

The other big change is the replacement of gradual, steady emission by sudden emission, the quantum lump of radiated energy—the photon—being the result of a quantum jump. (Bohr was *almost* there. He recognized the quantum jump and the suddenness of the radiation, but had not yet accepted that the light emerges as a photon.)

What, then, do line spectra reveal about atoms? That every atom has its own characteristic set of energy levels and that quantum jumps between these levels occur whenever light is emitted or absorbed.

Only one element, hydrogen, has a line spectrum so simple that it can be accurately correlated with theoretical values of the atomic energy levels. Reportedly, when Bohr had drafted his 1913 paper on atomic structure, he told his mentor Ernest Rutherford that he was reluctant to publish because he could explain only the spectrum of hydrogen, not of any other element. Rutherford is said to have responded, "If you can explain hydrogen, people will believe the rest." Indeed, the revolutionary ideas introduced by Bohr have stood the test of time and apply to all atoms, even if simple calculations are out of reach for the rest.

The hydrogen spectrum that Bohr was able to explain is called the *Balmer series*, named for Johann Balmer, a Swiss school teacher who, in 1885, recognized a simple numerical regularity in the inverse wavelengths of the hydrogen spectrum. Since frequency is inversely proportional to wavelength, his formula equally shows regularities in the frequencies, and therefore, as we now know, in the energy differences between the stationary states in hydrogen. Bohr used this formula. It is what he told Rutherford he could explain. Figure 20 shows a chart of the hydrogen-atom energy levels and the quantum jumps that give rise to the Balmer series.

30. Why is the chart of the elements periodic? The periodic table of the elements is the truly amazing result of the exclusion principle. To see how it works, imagine that you are Zeus and want to busy yourself making the world. You have one chest containing all possible atomic nuclei and another chest full of electrons. To get started, you pull out of the first chest the lightest nucleus, a single proton, and you bring an electron near it. The negatively charged electron is electrically attracted to the positively charged proton. After a while and after emitting a number of photons, the electron cascades down through the allowed quantum states of motion and arrives at the ground state, from which no further quantum jumps are possible. There it has zero orbital angular momentum and ½

FIGURE 20

In the hydrogen atom,
quantum jumps from
higher-energy states to
the second state give
rise to the Balmer series.

unit of spin, and its wave spreads over a size of about a tenth of a nanometer. The electron's energy is −13.6 eV, meaning that it would take that much energy to free it from the proton. The quantity 13.6 eV is called its *binding energy*. You, Zeus, have created element number 1, hydrogen.

Then you pull out a nucleus with 2 units of charge, one containing two protons—and, incidentally, two neutrons as well. You add an electron, which, just as before, cascades down to reach a ground state. It is much like the ground state of hydrogen except that its binding energy is four times as great, about 54 eV. But this combination still has a net positive charge. It is an ion, not an atom. So you reach for another electron and add it. It, too, cascades downward in energy to join the first electron in the lowest state of motion. Now enters the exclusion principle. Since you, as the god of all that is, want an interesting world, not a drab one, you have decreed that no two electrons can occupy the same state of motion at the same time.

But these two electrons have an escape hatch. They seem to be in the same state of motion, but they aren't, really. Their spins point in opposite directions, making for two different states of motion even when their orbital motions are identical. The helium atom that you have created has zero orbital angular momentum, and also zero spin, since the two electron spins point in opposite directions, and it has quite a large binding energy, about 79 eV (the energy needed to strip away both its electrons). It is smaller than the hydrogen atom. What you have created is an atom that is very unlikely to combine chemically with any other atom because it is so small and so tightly bound.

Warming to your task, you toss three electrons at a nucleus containing three protons (and three or four neutrons). Two of these electrons mimic the behavior of the two electrons in the helium atom. They pair up in a low-energy state with oppositely directed spins. But the third electron can't join them. The exclusion principle won't allow it. The third electron ends up in a state of motion of considerably greater energy (and, accordingly, much less binding energy) than the first two, and its wave, with an extra cycle of oscillation, spreads over a greater dimension. This third electron can have either zero or 1 unit of orbital momentum. You have created a lithium atom, much larger than a helium atom and much more loosely bound. This third element, lithium, will prove to be chemically quite active. Its outermost electron, called a *valence electron*, can participate readily in exchanges with other atoms. Here, already, is evidence that the exclusion principle makes the world "interesting," not "drab." Without that principle, all three electrons in the lithium atom would fall into a tightly bound state, and lithium would be even more reluctant than helium to interact with other atoms. Had you not decreed the exclusion principle, atoms of increasing number of protons and electrons would be more and more tightly bound, less and less inclined to react chemically. It would be a dismal world.

To address periodicity of the table of elements, we need to follow your role as Zeus, the atom maker, a bit farther. Consider what happens as you construct elements 4, 5, 6, 7, 8, 9, and 10. Because that third electron in lithium can have either zero or 1 unit of orbital angular momentum and because it has ½ unit of spin, it actually has a total of eight states of motion available to it. With zero orbital angular momentum, there are two possible states because there are two possible directions for the spin. With 1 unit of orbital angular momentum, there are three possible orientations of the orbital angular momentum, and, for each of them, two possible orientations of spin, for six more states—in total, eight states. This means that as you pick nuclei of successively greater charge and add more electrons, you can add a grand total of eight beyond what you had in helium, bringing you to element number 10, which is neon, and which, like helium, has a tightly bound set of electrons and little affinity for chemical activity. The two electrons in helium are said to occupy the first "shell." The outermost electrons in elements 3 (lithium) through 10 (neon) are said to be in the second shell.

One more step will make the entire process clear. What happens when you select a nucleus with eleven protons and add eleven electrons to it? The first two electrons fall into the first shell. The next eight fall into the second shell. Electron number eleven is excluded from those shells. It has to occupy a higher-energy, less tightly bound state, which is the beginning of the third shell. You have created another chemically active element, sodium, and it finds a place in the periodic table right under chemically active lithium, while the nearly inert number 10, neon, sits beneath the inert number 2, helium.

The process gets a bit more complicated as you go on—for instance, some shells hold eighteen instead of eight electrons—but basically the entire periodic table is shaped by the exclusion principle and the rules for orientation of angular momentum. Here is a question to wrestle with: If the number 8 comes from the available number of states with zero or 1 unit of orbital angular momentum, where does the number 18 come from?

31. Why are heavy atoms nearly the same size as lightweight atoms? As you move through the periodic table, there are fluctuations in the sizes of atoms. Going from lithium (number 3), for example, to neon (number 10), atoms get smaller, only to jump to larger size at sodium (number 11). But overall, from one end of the periodic table to the other, atoms are all of roughly the same size. The uranium atom, with its ninety-two electrons, is scarcely any larger than the beryllium atom, with its four.

Before explaining the reason for the puzzling uniformity of atomic size, let me discuss the simplest of atoms, the hydrogen atom. It has a certain size when the electron is in its lowest-energy state of motion (its ground state) and a larger size when the electron is excited to a higher-energy state. In fact, the size grows rapidly as the excitation energy increases. A hydrogen atom in its first excited state (principal quantum number $n = 2$) is four times as large in diameter as it is in its ground state ($n = 1$). In the second excited state ($n = 3$), its diameter is nine times as great as in the ground state. Hydrogen atoms have been identified in the lab with quantum number n greater than 40. These gargantuan atoms, more than 1,600 times normal size, are rare birds because even in a dilute gas they don't have enough elbow room to survive encounters with other atoms.

Because of the exclusion principle—the same principle that accounts for the periodicity of the elements—electrons for heavier elements get forced into states of higher and higher quantum number n (higher and higher shells). For the uranium atom, the outermost electrons have $n = 7$. Why, then, is the uranium atom not forty-nine times larger than the hydrogen atom? Because the uranium nucleus, with its ninety-two protons, pulls so hard on the electrons that their orbits are drawn in to a size comparable to that of the very light atoms.

The two innermost electrons in the uranium atom, those corresponding to the ground-state electrons in the helium atom, are pulled with forty-six times as much force as they experience in helium. As a result, the first shell in uranium is only one-forty-sixth the size of a helium atom. Other electrons pile up in shells of greater size until the outermost electrons are about as far from the nucleus as the few electrons in the light atom are.

32. How do protons and neutrons move within a nucleus? Nuclear physics can be said to have been inaugurated by Ernest Rutherford with his 1911 experiments that revealed a very small positive core within atoms. What was in that core remained mysterious for twenty years. By 1932, the year the neutron was discovered, physicists knew the approximate size of nuclei and knew that in high-energy collisions nuclei could react with each other and produce new nuclei. Since the nucleus is positively charged, they assumed that it contained protons, and they assumed further, although with trepidation, that the nucleus also contained electrons, since otherwise the total mass and total charge of a nucleus could not be reconciled.

J. Hans Daniel Jensen (1907–1973). After World War II, Jensen, with Heisenberg's support, was able to convince the authorities that he had joined the Nazi party in the 1930s because of expediency, not commitment. As a postwar professor at the University of Heidelberg, he helped keep theoretical physics alive and active in Germany. In villages outside of Heidelberg, Jensen knew where to find the best Obstschnaps (fruit liqueur). Photo courtesy of AIP Emilio Segrè Visual Archives, *Physics Today* Collection.

As I discussed at Question 8, James Chadwick's discovery of the neutron made it clear at once that the nucleus need not contain electrons. The nucleus can be understood as a collection of protons and neutrons. Physicists could breathe a sigh of relief once they banished electrons from the nucleus, but they were left with the question of how the protons and neutrons (the "nucleons") organize themselves within the nucleus. Do they behave more like a solid, a liquid, or a gas? As I reported at Question 8, for a decade or so the liquid droplet model worked well, and, in particular, it did a good job of accounting for nuclear fission, discovered at the end of 1938 and explained theoretically in 1939.

After World War II the liquid droplet model began to show some imperfections. Physicists were seeing evidence that protons and neutrons within the nucleus occupy shells, just as electrons do in atoms. This could mean only that these nuclear particles move somewhat freely within the nucleus. Nuclei with particular numbers of protons or neutrons showed unusual stability, just as atoms with particular numbers of electrons do. For atoms, the closed-shell numbers are 2, 10, 18, 36, 54, and 86. These are the numbers of electrons in the atoms of the so-called noble gases: helium, neon, argon, krypton, xenon, and radon. In each of these atoms a shell is filled, leaving no room in that shell for an additional electron. The closed-shell numbers in nuclei are not the same, and

Maria Goeppert Mayer (1906–1972), shown here with her daughter Marianne in 1935. Born and raised in Germany, Mayer met and married the American chemist Joseph Mayer in Göttingen, where she was a student. She was the seventh generation in a row on her father's side to become a university professor, although for her it was no easy matter. Fortunately, she loved physics and pursued it for years without any prestigious title and sometimes even without a salary. Photo courtesy of AIP Emilio Segrè Visual Archives, Maria Stein Collection.

this caused some initial perplexity. They are 2, 8, 20, 28, 50, 82, and 126. In fact, the nuclear closed-shell numbers were at first called the *magic numbers*, because physicists could not figure out why they were what they were. The physicists who independently provided the explanation were Maria Goeppert Mayer* in America and a team led by J. Hans Jensen† in

---

* For many years Maria Mayer had to make do with part-time or temporary appointments while her husband, the chemist Joe Mayer, enjoyed professorships. When I visited her to discuss nuclear physics in the 1950s, she was holding two half-time positions, one at Argonne National Laboratory and one at the University of Chicago. (This was after her pathbreaking work on nuclear shell structure.) Eventually both she and her husband were appointed as professors at the University of California, San Diego.

† Hans Jensen was a bachelor who lived in an apartment in the physics building at the University of Heidelberg. Once when I visited him, he told me excitedly that he had just returned from America, where he had learned to mix a martini. He offered to show me his new skill and I accepted. He then combined three parts vermouth and one part gin and served it warm. Maria Mayer was more of a pro in this department.

Germany. They (together with the Hungarian-American physicist Eugene Wigner for unrelated work in nuclear physics) shared the 1963 Nobel Prize in physics for their insight. They recognized that the exclusion principle is the basic reason for shell structure in nuclei just as it is in atoms, but that somewhat different rules for combining spin and orbital angular momentum are at work in nuclei.

Because the "magic numbers" (closed shell numbers) apply to protons and neutrons separately, it is possible for a nucleus to be "doubly magic"— that is, to have both its neutrons and its protons filling shells. The lightest doubly magic nucleus is that of helium-4, containing two protons and two neutrons. It is such a tightly bound structure that it can be emitted as a unit—namely, an alpha particle—from a radioactive nucleus, and, for the same reason, it is widely distributed in the universe. The heaviest doubly magic nucleus is that of lead-208, with 82 protons and 126 neutrons. This isotope is the heaviest stable nucleus found in nature.

How can it be that a nucleus can seem to be a liquid droplet and at the same time act like a gas of free nucleons? The answer lies in the nature of the nuclear force (which, ultimately, arises from the exchange of gluons among quarks). This force allows a proton or neutron to glide more or less unimpeded from one side of a nucleus to the other. But if a nucleon tends to stray away from its mass of fellow nucleons, it is pulled back sharply into the fold. This strong force at the edges of the nucleus produces something much like the surface tension of a liquid. So the nucleus as a whole can vibrate and oscillate like a liquid droplet even while the particles within it move more like the molecules in a gas. The theory of this dual nature of a nucleus goes by the name *unified model* or *collective model*. Niels Bohr's son Aage (pronounced OH-uh), together with his American colleague Ben Mottelson* won the 1975 Nobel Prize in Physics for their work on the unified model.

---

* By chance I came to know Ben Mottelson in 1944 when we were both high school seniors headed toward careers in physics. He settled in Copenhagen, where English was (and is) the principal language in the Bohr Institute. When I visited him there in the 1950s, his children explained to me, "We speak Danish when we don't want Mommy and Daddy to understand."

33. What are atomic number and atomic mass? I will define symbols as I define the quantities. The total number of protons in a nucleus is denoted by Z and is called the *atomic number*: $Z = 1$ for hydrogen, 2 for helium, 10 for neon, 26 for iron, 92 for uranium, and so on. Evidently Z also labels the element in the periodic table. The element with the largest value of Z—the largest atomic number—detected so far is element number 118, with, to date, no official name.*

The number of neutrons in the nucleus is denoted by N and is called, straightforwardly, the *neutron number*. The sum of $Z + N$, the total number of nucleons in the nucleus, is denoted by A and is called the *mass number*. Thus the mass number A of the most common isotope of hydrogen is 1 ($Z = 1$, $N = 0$), and hydrogen's heavier isotopes, deuterium and tritium, have mass numbers 2 and 3, respectively ($Z = 1$, $N = 1$ for deuterium; $Z = 1$, $N = 2$ for tritium). The most common isotope of carbon is carbon-12, with $Z = N = 6$ and $A = 12$. Jumping to near the other end of the periodic table, uranium-238 has $Z = 92$, $N = 146$, and $A = 238$.

There is one more related quantity, which is the actual mass M of the nucleus. It turns out to be much more convenient to "weigh" atoms, including their electrons, than to weigh bare nuclei, so masses are actually specified for whole atoms, including the electrons. Electrons, in any case, make up a very small part of an atom's mass. The large-scale mass unit, the kilogram, is miles away from being useful to designate atomic masses, so a new unit, called the *atomic mass unit*, or *amu*, has been introduced.†️ It is defined in such a way that an atom of carbon-12 has an atomic mass of *exactly* 12 amu. With this definition, the atomic masses of the hydrogen isotopes are close to 1, 2, and 3; the atomic mass of oxygen-16 is close to 16; and the atomic mass of uranium-238 is close to 238. In other words, the integer A, which is the mass number, is roughly equal to the actual mass M of an atom expressed in amu. The atomic mass of oxygen-16, to give an example, is 15.9949, and that of beryllium-9 is 9.0122. (Both are known to even greater precision than indicated here.) The reason that

---

*Its placeholder name is *ununoctium* (a Latinized version of *one-one-eight*).
† 1 amu is equal to 1.66×10^{-27} kg.

atomic masses and mass numbers are not *exactly* the same is that nuclear binding energy contributes to nuclear mass.

I have used the word *isotope* without carefully defining it. First, an *element* is defined as a material whose atoms all contain a particular number of protons. In other words, the atomic number Z defines the element. A given element may have different isotopes, depending on the number of neutrons in its nuclei. The two famous isotopes of uranium, for instance, are U-235 and U-238. All uranium atoms contain ninety-two protons. The atoms of the two isotopes contain, respectively, 143 and 146 neutrons. The most common isotope of carbon is carbon-12, whose nucleus contains six protons and six neutrons. The rarer isotope carbon-14, used for dating ancient artifacts, has a nucleus containing six protons and eight neutrons. It is radioactive, whereas the nucleus of carbon-12 is not.

Since the number of electrons in a neutral atom and the way in which they are arranged depend only on the number of protons in the nucleus, the behavior of atoms in bulk depends hardly at all on the neutron number. Thus, carbon-12 and carbon-14 enter into identical chemical compounds, and both would be equally suitable for making graphite or diamonds. That is why it is only the number of protons that defines the element. (Chemical behavior is not *totally* independent of neutron number, since the mass differences among isotopes can play a small role.) When the nucleus itself is viewed as a separate structure, however, isotopic differences can loom large. The nucleus of lead-209, for example, with its one "valence" neutron, has properties quite different from those of the lead-208 nucleus. The most pronounced difference among isotopes may be whether they are radioactive or not, as for carbon-12 and carbon-14. Most elements have more than one isotope. Eighty of the elements have at least one stable isotope, and *every* element has one or more unstable (radioactive) nuclei. Among the known elements, more than three thousand isotopes have been identified, of which 256 are stable. The nuclear physicist, focused on these atomic cores, can ignore the "remote" electrons and the chemistry of atoms in the same way that a terrestrial geologist can ignore the solar system and the galaxy.

section VI

And More about Nuclei

34. Why does the periodic table end? The ending of the periodic table (at $Z = 118$ for all currently known elements, at $Z = 82$ for stable elements*) has nothing to do with electrons and everything to do with nuclei—and, specifically, everything to do with the electric repulsion between protons. The strong nuclear force attracts protons to each other, neutrons to each other, and neutrons to protons. In short, it acts to bind all nucleons together. It doesn't reach out very far, not even as far as from one side to the other side of a large nucleus. It acts only when nucleons are close together. The electric repulsion between protons is weaker than the nuclear force but of longer range. In a heavy nucleus, every proton feels the repulsive force of all the other protons.

The exclusion principle acts among protons and among neutrons but not between protons and neutrons. So no two protons can occupy the same state of motion within the nucleus, and no two neutrons can occupy the same state of motion, but a proton and a neutron have no such bar and can share a state of motion. This means that if there were no electric force to get in the way, the most stable nuclei would all have equal numbers of neutrons and protons. The neutrons and protons would

---

*Elements number 43 and 61 (technetium and promethium) have no stable isotopes. That is why there are eighty stable elements, even though the last one is number 82.

pile up in their respective shells, each obeying the exclusion principle but intermingling freely with each other. A stable iron nucleus would contain twenty-six protons and twenty-six neutrons. A stable uranium nucleus would contain ninety-two protons and ninety-two neutrons. And there would be no limit to how many neutrons and protons could join together to form ever larger nuclei. Element number 548 would have a nucleus containing 548 protons and 548 neutrons. The international panel that names elements would have its hands full.

Now to the real world, in which protons do repel one another. Their mutual repulsion is relatively unimportant for the lighter nuclei, where indeed equal numbers of protons and neutrons predominate—for example, in helium-4, with two protons and two neutrons; in oxygen-16, with eight protons and eight neutrons; and in neon-20, with ten of each. But as the number of protons gets larger, their mutual repulsion becomes relatively more important, overcoming the tendency toward equal numbers of neutrons and protons. For the heavier nuclei, neutrons outnumber protons, and the heavier the nucleus, the more the imbalance. In the most common iron isotope, iron-56, there are thirty neutrons—15 percent more than the twenty-six protons. In uranium-238, an isotope I have mentioned before, there are 146 neutrons, 59 percent more than the ninety-two protons. Eventually, the repulsive force among the protons is so great that there are no stable nuclei at all. And with no stable nuclei, there are no long-lasting atoms and no elements for scientists to study.

If a graph is constructed in which the neutron number N is plotted horizontally and the proton number Z is plotted vertically (see Figure 21), a line can be drawn representing the stable nuclei. In the hypothetical world in which protons don't repel each other, that line, the "line of stability," is a straight line at an angle of 45 degrees, since then neutron and proton number are equal $(N = Z)$ for the most stable nuclei. This is indicated by a dashed line in the figure. In the real world, proton repulsion causes the line of stability, shown by the solid line in the figure, to *bend* and to *end*. It bends because the neutron number N gets increasingly large relative to the proton number Z. It ends because beyond a certain point there is no way, even with the help of neutrons, to pack more protons into a nucleus.

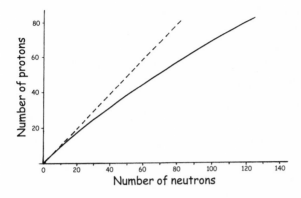

FIGURE 21
Number of protons
vs. number of
 neutrons in stable
nuclei. This "line of
stability" bends and
ends.

Neutrons, it turns out, not only help extend the line of stability through their role in keeping protons at arm's length and by adding extra cement within nuclei. They are also, in a sense, responsible for the very existence of nuclei. Without the neutron there would be no helium or any other heavier element. There would be a universe consisting only of hydrogen, with no planets and no people. This is because two protons, by themselves, can't stick together to form a nucleus. There is no such thing as a nucleus of helium-2 (two protons and no neutrons). Despite the relative weakness of the electric repulsion, it is strong enough to prevent the bonding together of two protons by themselves. Only if they are joined by one or two neutrons can they hang together. The nuclei of helium-3 (two protons and one neutron) and helium-4 (two of each) are stable, and, in a sense, provide a platform for the construction of all heavier nuclei.

Think of the marvels of the world that follow from a few simple principles and facts. Without the neutron there would be no element other than hydrogen. Without the exclusion principle and the rules for combining angular momenta, there would be a drab, largely inert set of atoms, no periodic table, and a world without life and color. Without the proton's charge, there would be no assembling of electrons into atoms, and nuclei would exist in endless number with nothing useful to do.

35. What is radioactivity? What are its forms? When a nucleus is unstable, it transforms itself (with a certain half-life) into something else that is more stable. That is radioactivity. The process got its name before the existence of the nucleus was known. Back at the end of the nineteenth century, scientists such as Ernest Rutherford and Marie Curie discovered that certain heavy elements spontaneously emit "radiation." At first lacking any knowledge of the detailed nature of the radiation, Rutherford named the first two kinds to be discovered *alpha* and *beta*—the first two letters of the Greek alphabet. Soon afterward, Paul Ulrich Villard, a French chemist, discovered a third kind of radiation, which he logically named *gamma*. Scientists soon pinned down the nature of these "rays." Alpha rays are helium nuclei—two protons and two neutrons bonded together—which we still call *alpha particles*. Beta rays are electrons, which, in some contexts, we still call *beta particles*. And gamma rays are electromagnetic radiation, which we recognize now as high-energy photons (and which we still call *gammas*).

Alpha and beta decay involve "transmutation"—that is, the initial and final nuclei belong to different elements. Gamma decay, on the other hand, leaves the element unchanged. It is simply a quantum jump from a higher to a lower energy state in the nucleus—quite like quantum jumps of electrons in atoms, but with much greater energy, as much as a million times greater.

Marie Curie (1867–1934), deservedly one of the most famous women scientists of all time and one of only three persons ever to win two Nobel Prizes in science (hers in physics in 1903 and chemistry in 1911). She and her husband Pierre discovered the elements radium (atomic number 88) and polonium (number 84), the latter named for her native Poland. She herself is honored in the name of the element curium (number 96). Deutscher Verlag; photo courtesy of AIP Emilio Segrè Visual Archives, Brittle Books Collection.

Why does radioactivity occur? What does it accomplish? Let me discuss alpha and beta decay separately.

Alpha decay occurs almost exclusively for very heavy elements, number 83 and beyond. It is really the nucleus's way of shedding protons, for it occurs in that part of the periodic table where proton repulsion within nuclei makes the nuclei barely stable, or even unstable. Now $E = mc^2$ gets into the act in a very interesting way. The nucleus with too many protons would "like" to get rid of a proton. But it can't do so. A nucleus has less mass than the sum of the masses of the particles within it. This is because the attractive nuclear force creates a binding energy. In effect, the nucleons within the nucleus, because of their attraction for each other, have less mass than if they were free. So if a proton is ejected, the nucleus doing the ejecting has to come up with some extra energy to restore the proton to its mass in the outside world. This stops the process. The same consideration bars the ejection of a single neutron or a pair of protons. But all is not lost. The nucleus has a way of getting rid of protons. It can eject two of them tightly bound with two neutrons—that is, an alpha particle—because the nucleons in the alpha particle have, effectively, less mass than if they were free. The nucleus doesn't have to come up with extra energy to make possible the ejection of an alpha particle. What it shoots forth is a tightly bound structure with considerable binding energy of its own. The neutrons play a helping role. The nucleus has no incentive (if I may anthropomorphize a bit) to get rid of neutrons, but it has to accept the loss of neutrons to achieve its objective of shedding protons.

The only thing that prevents many heavy nuclei from undergoing alpha decay with extreme rapidity is an electric force barrier. It would take great energy for an alpha particle to sneak just a tiny bit away from the nuclear surface, for it is drawn back by the nuclear force. To make its escape, it must "tunnel" through the barrier to emerge at some greater distance from the nuclear surface, from where it can fly free. As I mentioned at Question 26, physicists in Britain and America developed the theory of alpha decay as a tunneling phenomenon in 1928. It all depends on probability. Classically, alpha decay cannot occur at all. Quantum mechanically, it can occur with some probability. It is, in many instances, a very small probability. The alpha-decay half-life of uranium-238, for instance, is 4.5 billion years, coincidentally about the same as the age of Earth.

A well-known alpha emitter with a shorter half-life is radium-226. Its decay into radon-222 can be displayed as a reaction equation:

$$_{88}\text{Ra}_{138}^{226} \rightarrow {}_{86}\text{Rn}_{136}^{222} + \alpha.$$

The subscripts to the left of the chemical symbols show the atomic number—that is, the number of protons in the nucleus. The subscripts to the right of the symbols show the number of neutrons. The superscript is the sum of these two, the mass number—that is, the total number of nucleons in the nucleus. The half-life for this process is about 1,600 years. The product of the decay, radon-222, is itself an alpha emitter, with a half-life of 3.8 days. The process shown here is part of a long chain of radioactive decay that begins with uranium-238 and ends with lead-206.

The physics for beta decay is quite different, and examples of beta decay are far more numerous. The hundreds of known beta emitters range from hydrogen-3 (tritium) at one end of the periodic table to the very heaviest elements, the "transuranics," at the other end. Beta decay is the result of the weak interaction, experienced by electrons and neutrinos as well as by neutrons and protons. This interaction triggers radioactivity whenever the numbers of protons and neutrons within a nucleus are out of balance. I can best explain what I mean by *out of balance* with an example. Consider a nucleus containing a total of fourteen nucleons. The most stable nucleus with mass number 14 is nitrogen-14, whose nucleus contains seven protons and seven neutrons. As is typical for the lightest elements, an equal number of protons and neutrons is favored. Another nucleus with the same mass number is that of carbon-14, with six protons and eight neutrons. Its complement of protons and neutrons is "out of balance" in that a more stable nucleus can be formed if one of carbon-14's neutrons is replaced by a proton. Thanks to the weak interaction, exactly this can happen. In the beta decay of carbon-14, an electron and an antineutrino are emitted and a neutron changes into a proton. Here is the process, displayed as a reaction equation:

$$_{6}\text{C}_{8}^{14} \rightarrow {}_{7}\text{N}_{7}^{14} + \text{e}^{-} + \bar{\nu}.$$

The Greek letter ν (nu) stands for the neutrino, and the bar over the symbol designates the antiparticle. The half-life for the process is 5,730 years, making it an excellent tool for radioactive dating of ancient (but not too ancient) artifacts. In the atmosphere, carbon-14 is replenished, but not, for example, in a tree after it is cut down. So the depletion of carbon-14 in the wooden artifact tells how long it has been since that wood was part of a living tree.

Neutrons and protons can be out of balance in the other direction as well. Nuclei of atomic mass 18 provide an example. At this mass number, a slight neutron excess is preferred, so oxygen-18, with eight protons and ten neutrons, is somewhat more stable than fluorine-18, with nine protons and nine neutrons. Fluorine-18 transforms itself (with a 110-minute half-life) into oxygen-18 as follows:

$$_9F_9{}^{18} \rightarrow {}_8O_{10}{}^{18} + e^+ + \nu.$$

A positron (antielectron), along with a neutrino, is emitted.

There is a third kind of beta decay, which also takes place when a nucleus can be made more stable by replacing a proton with a neutron. This is called *electron capture*. Instead of emitting a positron, as fluorine-18 does, the nucleus may gobble up an electron from within the atom. (Note that in both cases the nuclear charge decreases.) This occurs, for example, when argon-37 transforms itself into chlorine-37:

$$_{18}Ar_{19}{}^{37} + e^- \rightarrow {}_{17}Cl_{20}{}^{37} + \nu.$$

Here at mass number 37, where the repulsion among protons within the nucleus is beginning to become important, the excess of three neutrons in chlorine-37 is favored over the excess of only one neutron in argon-37.* (Following this process, the chlorine-37 atom finds itself missing one

---

*In the 1970s Raymond Davis used the reverse of this reaction to detect neutrinos from the Sun for the first time (an achievement that earned him the 2002 Nobel Prize in Physics). His one-hundred-thousand-gallon tank of dry-cleaning fluid (perchloroethylene) in a deep mine in South Dakota contained lots of chlorine-37. After waiting a couple of months so that some of the rare neutrino-absorption

inner electron. The resulting emission of an X-ray by the atom is what makes the detection of this process possible.)

36. Why is the neutron stable within a nucleus but unstable when alone? In one respect, at least, the neutron is the most remarkable of all particles. Alone it decays with a mean life of about fifteen minutes; within a nucleus it can live forever.* It is like an animal that cannot survive in the wild but prospers when caged in a zoo.

As with so much else in the nuclear world, it's all about mass and energy. In energy units, the mass of a neutron is 939.6 MeV and the mass of a proton is 938.3 MeV, less by 1.3 MeV, or about 0.1 percent. This small difference is enough to permit the decay of the neutron, which is represented by the reaction equation

$$n \rightarrow p + e^- + \bar{\nu}.$$

The electron's mass soaks up an energy of 0.5 MeV, and the antineutrino's mass a much smaller amount, so 0.8 MeV is left over in this "downhill" decay to power the recoil of the three particles that are created.

Now let's join the neutron to a proton to form a deuteron, the simplest nucleus other than a lone proton. The mutual embrace of these two particles is sealed by a binding energy of 2.2 MeV. This means that 2.2 MeV is released when the deuteron is formed, and, conversely, 2.2 MeV must be added to the deuteron to break it apart into a neutron and a proton. It also means (thanks to $E = mc^2$) that the deuteron has less mass than the sum of the masses of its two constituent particles. And why can't the neutron in the deuteron decay as it does "in the wild" to turn the deuteron into a pair of protons (plus an electron and an antineutrino)?

---

events would occur, he swept from the fluid the tiny amount of argon-37 that had accumulated, and detected this isotope by its radioactive transformation back to chlorine-37 with a half-life of thirty-four days.

* There are radioactive nuclei whose decay involves the disappearance of a neutron, but, strictly speaking, in such instances it is the entire nucleus, not any neutron within it, that decays.

Because for it to do so would be "uphill" in mass. Let's do the arithmetic (in MeV units):

Mass of proton	938.3
Mass of neutron	939.6
Less binding energy	− 2.2
Mass of deuteron	**1875.7**
Mass of two protons	1876.6
Mass of electron	0.5
Mass of antineutrino	~0.0
Mass of products of hypothetical decay	**1877.1**

The products of the hypothetical decay weigh in at 1877.1 MeV, more than the deuteron's mass, so the decay does not take place.

I have skipped over one important consideration. You might wonder whether the two protons resulting from a deuteron decay could stick together to form a "di-proton,"* whose binding energy would lower the second total above and make the decay real, not hypothetical. As I mentioned at Question 34, there is no stable di-proton. The electric repulsion between a pair of protons outcompetes their nuclear attraction and they fly apart. So, in fact, the deuteron can't decay. The neutron within it is stabilized and lives forever. Indeed, the universe now undoubtedly contains countless deuterons that were created billions of years ago, not long after the Big Bang. Their neutrons, wrapped in the embrace of protons, last through the ages.

As it turns out, there is no di-neutron either. Two neutrons, despite the absence of electric repulsion between them, don't stick together. The reason is that the nuclear force between a pair of nucleons is just a little stronger if their spins point in the same direction than if they are oppositely directed. For a neutron and a proton, this is permissible. Their spins are aligned in the deuteron, which has a total spin of 1 ($\frac{1}{2} + \frac{1}{2}$).

---

* If it existed, the di-proton would be the nucleus of the lightest isotope of helium.

But the exclusion principle prohibits a pair of neutrons in their lowest-energy state from having identical spin directions. They have to combine with opposite spins, in which arrangement the force between them is not quite sufficient to hold them together. The deuteron rests on a knife edge of stability.

For heavier nuclei, although the changes in mass produced by binding energy are still small, they are quite a bit larger than for the deuteron, making the neutron's stabilization more "robust" in heavier nuclei. The binding energy of an alpha particle, for instance, is 28 MeV, or 7 MeV per nucleon, much more that the deuteron's 1 MeV per nucleon. Beyond the alpha particle (a helium nucleus), the binding energy per nucleon climbs to about 8 MeV for carbon and oxygen nuclei, reaches a peak of about 9 MeV around iron and nickel, and falls gradually back to about 7 MeV for the heaviest nuclei.

Without the stabilization of the neutron, it would be a dull universe indeed, consisting only of hydrogen—and cold hydrogen at that, as there would be no nuclear fusion to release energy and light the stars. It's a sobering thought that if the neutron were a little more massive than it actually is or the nuclear force a little weaker, there would be no stabilization of the neutron and no us.

37. What is nuclear fission? Why does it release energy? The competition within the nucleus between attractive nuclear forces and repulsive electric forces goes a long way toward explaining the properties and behavior of nuclei. As I've already discussed, it explains why protons and neutrons cluster in equal number in light nuclei, why neutrons outnumber protons in heavy nuclei, why alpha and beta radioactivity occur, and why the periodic table ends. There is one more important thing that it explains: why fission of heavy nuclei and fusion of light nuclei release energy. *Fission* and *fusion* mean pretty much what the names suggest. Nuclear fission is the breaking apart of a heavy nucleus into lighter fragments. Nuclear fusion (addressed in the next question) is the joining together of light nuclei to form a heavier nucleus.

To understand fission and fusion, it helps to grapple with the concept of mass per nucleon. As mentioned at Question 36, the binding energy for most nuclei ranges between 7 MeV and 9 MeV per nucleon (although it is only 1 MeV per nucleon for the deuteron). The effect of this binding energy (thanks to $E = mc^2$) is to reduce the nuclear mass. It's as if each nucleon, when joined to others, has slimmed down a bit. In a commercial for togetherness, a neutron could say, "I used to weigh 939 MeV, but now, after joining with my friends, I weigh only 931 MeV." To see how the average mass per nucleon is calculated, you could think of, say, eight protons and eight neutrons, well separated from one another. They have a certain total mass. Bring them together to form an oxygen nucleus and they have a smaller total mass. Divide that mass by sixteen and you have the mass per nucleon of the oxygen nucleus. If you divide the mass of a brick wall by the number of bricks in it, you get the mass of a single brick. If you divide the mass of a nucleus by its number of "bricks" (nucleons), you get something smaller than the mass of an isolated single nuclear "brick."

Figure 22 shows the average mass per nucleon for elements ranging from hydrogen to uranium. Its vertical scale is selective to accentuate the relatively small changes that occur from one end of the periodic table to the other. From the sharp peak at the left (a lone proton, not slimmed at all) to the bottom of the "mass valley" around element 26 (iron) is a change of less than 1 percent. From that same valley up the

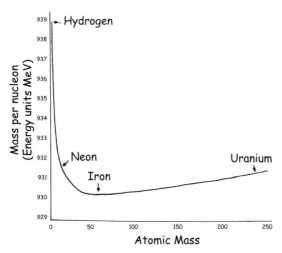

FIGURE 22
For nuclei, the mass per "brick" (nucleon) is not constant. It falls to a minimum value at iron and rises for heavier nuclei.

slope to element 92 (uranium) on the right is a change of 0.2 percent. Small though these changes might be in terms of percentage, they are momentous. Because of the left-to-right downhill slide on the left, the stars shine, hydrogen bombs explode, and controlled fusion power might someday come to humankind. Because of the right-to-left downhill slide on the right, atomic bombs explode and nuclear power stations provide countless megawatts of electric power.

Consider, first, fission. If a nucleus toward the right side of the graph in Figure 22 splits into two smaller pieces, each of those pieces will be farther to the left in the graph and will have less mass per nucleon. The transformation will be downhill in mass as well as literally downhill in the figure. Energy will be released. Then the question is "Why doesn't it just happen, and happen quickly?" In fact, for some of the heaviest nuclei, spontaneous fission does occur, but not quickly. Half lives are typically years or millennia or longer. The nucleus has to tunnel through an energy barrier—a process of very low probability—either to emit an alpha particle or to break into more nearly equal fragments. Among elements beyond uranium in the periodic table—all of which are radioactive—alpha decay is common. For a few, spontaneous fission is a principal mode of decay.

There is one thing that can happen to enable some nuclei to vault over the energy barrier instead of tunneling through it. Consider a uranium-235 nucleus as an example. Left to itself, it can undergo spontaneous fission, with a half-life of 100 million billion years.* If it gains 5 MeV or more of energy, it surmounts the energy barrier and undergoes fission in something like a billionth of a billionth of a second. One way for it to acquire that energy is to absorb a neutron. When an approaching neutron is "pulled" into the nucleus, a binding energy of about 5 MeV is released, which goes into exciting the resulting uranium-236 nucleus. This excited nucleus doesn't have to tunnel. It can break apart almost instan-

---

* This would be its half-life if spontaneous fission were its only mode of decay. Because its predominant mode, alpha decay, is much more likely, its actual half-life is "only" 700 million years. Thanks to the working of probability, however, spontaneous-fission events can be observed in U-235 in minutes or hours or days.

taneously, and when it does so there is a leap downward from right to left on the graph in Figure 22. This leap to lower-mass fragments releases more than 200 MeV of energy.

From a human perspective, even 200 MeV would not amount to much were it not for a chain reaction. It turns out that when the fission process occurs, two or three neutrons are emitted along with the two large fission fragments. These neutrons can find other uranium-235 nuclei and cause additional fission events. When this chain-reaction process is carefully controlled within a reactor, circulating water can carry off the energy that is produced and electric power is the end result. When the chain-reaction process is uncontrolled, the rate of fission events grows exponentially and a nuclear explosion is the end result.

A word on why uranium-235 and not uranium-238 underlies fission in reactors and bombs. It is a subtle difference indeed. With its 146 neutrons, the U-238 nucleus is just a tiny bit larger than the U-235 nucleus, with its 143 neutrons. As a result, the 92 protons in U-238 are just a little farther apart on average than are the 92 protons in U-235, making the effect of proton repulsion ever so slightly less in U-238 than in U-235. The heavier isotope is therefore a bit more stable, and the 5 MeV that is sufficient to make a U-235 nucleus come "unglued" (that is, to surmount its energy barrier) is insufficient to do the same in U-238.* The explanation of this difference, along with a general theory of the fission process, was published by Niels Bohr and John Wheeler on September 1, 1939, the day that Nazi Germany invaded Poland and set off World War II (and only nine months after the discovery of fission).

In natural uranium found on Earth, about one atom out of 140 (0.7 percent) is the isotope U-235. This is an insufficient abundance to permit a chain reaction to take place in uranium ore.† But it was not always

---

* A second, smaller effect also comes into play. Because of an "odd-even" effect (ultimately a result of the exclusion principle), a neutron absorbed by U-235 provides a somewhat greater binding energy than one absorbed by U-238, and it helps get the nucleus up and over the energy barrier.

† In 1942 Enrico Fermi was able to coax a chain reaction from natural uranium, with its 0.7 percent U-235 content, but this required a carefully constructed "pile"

so. Uranium-238 decays with a half-life of about 4.5 billion years. Its lighter sibling U-235 decays at a much brisker rate, with a half-life of 700 million years. Two billion years ago, in a uranium deposit on Earth, about one atom in 27 (3.7 percent) was U-235. This could be enough to make possible a chain reaction in nature. In 1972, a French physicist, Francis Perrin, found evidence that indeed such a chain reaction had occurred in a uranium ore deposit in what is now Oklo, Gabon, nearly 2 billion years ago, and that this "natural reactor" had continued to churn out fission energy for more than one hundred thousand years.

38. What about nuclear fusion? Here is an example of nuclear fusion that is of practical importance for both thermonuclear weapons and potential future fusion reactors. A deuteron and a triton come together to form an alpha particle and a neutron:

$$d + t \rightarrow \alpha + n.$$

In a notation that makes clear the atomic number, neutron number, and mass number, this fusion reaction can be written as

$$_1H_1^2 + {}_1H_2^3 \rightarrow {}_2He_2^4 + {}_0n_1^1.$$

The two protons in the reacting hydrogen nuclei end up in the alpha particle. Of the three neutrons in the hydrogen nuclei, two end up in the alpha particle (the helium nucleus) and one is set free. It is the large binding energy of the alpha particle that makes the particles tumble downhill on the left side of the graph in Figure 22. The energy released in this reaction is more than 17 MeV, most of which (14 MeV) appears as kinetic energy of the final neutron.

The fusion fuel in the Sun is mostly nuclei of the lightest isotope of hydrogen—that is, simply protons. Through a series of reactions, protons in the Sun's core fuse to make alpha particles according to the following net reaction:

---

of uranium deployed with pure graphite, an arrangement that would never happen "in the wild." Enrichments in modern power and research reactors range from around 2 percent to around 20 percent.

$$4p + 2e \rightarrow \alpha + 2\nu.$$

Again, looking at the left side of Figure 22, you can see that this is a downhill reaction that releases energy (hydrogen to helium). To conserve charge and lepton number (which I discuss later, at Questions 45 and 55), two electrons must be gobbled up in this net reaction and two neutrinos must be produced (neutrinos that have now been detected on Earth, as mentioned in the footnote on page 91). In the notation that displays atomic number, neutron number, and mass number, the Sun's net fusion reaction is written

$$4_1H_0^1 + 2_{-1}e_0^0 \rightarrow {}_2He_2^4 + 2_0\nu_0^0.$$

If you look carefully at this reaction equation, you will see that the numbers of protons and neutrons are not the same before and after. In effect, two protons are changed into two neutrons. The net reaction involves the weak interaction, the same interaction that is responsible for beta decay. It's a good thing, in a way, that the weak interaction is a participant in the Sun's fusion reactions. That helps explain why the Sun has been shining already for about 5 billion years and will shine for 5 billion more. Each second, the Sun converts 4.6 million tons of mass into energy.

I still need to answer the same question about fusion that I asked about fission: If it releases energy, why doesn't it happen immediately? Why do protons mill about in the middle of the Sun for up to billions of years before fusing and releasing the energy that, in the form of extra mass, has been there all along? It seems to be like a child placed at the top of a slippery slide who just sits there for hours without moving and then suddenly starts to slide. For protons (indeed for any nuclei), the sluggishness is rooted in the electric repulsion between protons (or nuclei). If one proton is hurled at another with sufficient energy in an accelerator, they do fuse readily. But even at the millions of degrees of temperature at the center of the Sun, the protons move so slowly that they are far more likely to deflect away from each other than to approach close enough to undergo a nuclear reaction. There are just two ways in which they can join up and fuse to release energy. One way is for the occasional proton with far more kinetic energy than the average to scoot up close to another proton despite their repulsion. Another way involves tunneling. In a process that is in some ways the opposite of what happens in alpha decay, a proton can tunnel through the energy barrier created by the electric repulsion. The actual process of fusion in the Sun takes place in both of these ways, which, fortunately for us, are slow, slow, slow. To go back to the analogy of the child at the top of the slide, it's as if there is a little hump to be surmounted before the child can start sliding down. If the child jiggles about enough, eventually he or she will get over the hump and then start down. Or the child could turn quantum-mechanical and, with some small probability, tunnel through the hump in order to reach the downward sloping slide.

Fusion reactions in the Sun are called *thermonuclear* because they are nuclear reactions that are made possible by high temperature, the effect of which is to elevate the kinetic energy of the nuclei, making their interaction more likely. On Earth, too (except in isolated cases at accelerators), fusion occurs as the result of high temperature. A thermonuclear weapon, or so-called H-bomb, uses a fission bomb to achieve a temperature high enough to ignite fusion reactions—reactions that, once established, may be able to sustain themselves.* Major research programs are under way to achieve thermonuclear reactions that are controlled rather than explosive. These may power future fusion reactors.

---

*These reactions include the so-called d-t reaction cited on page 98.

Particles

*T*he questions in Sections V and VI moved down the scale of size through atoms to atomic nuclei. In this section and the next, I continue on down that scale to particles. Their names and their properties can come across as a confusing jumble, but they stand as ideal exemplars of the quantum world— first, because they are, at least for now, the ultimate bits of matter and energy that underlie all else, and second, because their properties and behavior illustrate particularly clearly how the governing principles of quantum physics work. That is why I address the particles now, before turning, in later sections, to the quantum principles they obey.

39. What is a lepton? What are its flavors? As set forth in Table A.1 in Appendix A, there are six leptons—three of them charged (the electron, muon, and tau) and three of them uncharged (the neutrinos that "belong," respectively, to the three charged leptons). From Thomson's discovery of the electron in 1897 until the discovery of the tau neutrino in 2000 by a group at Fermilab in Illinois, more than a century elapsed. As the following dates suggest, it was a painstaking process to get from the first lepton to the last.

  1897   J. J. Thomson discovers the electron.
  1930   Wolfgang Pauli postulates a neutrino.

1934 Enrico Fermi's theory of beta decay makes the neutrino believable.

1937 Cosmic-ray experiments show particles of intermediate mass ("mesons").

1947 Cecil Powell segregates the mu meson (later *muon*) from the pi meson (later *pion*).

1956 Frederick Reines and Clyde Cowan Jr. identify the electron neutrino.

1962 Brookhaven Lab group discovers the muon neutrino.

1978 Martin Perl discovers the tau.

2000 Fermilab group discovers the tau neutrino.

The three dates in the 1930s identify suggestive developments. The other dates are the discovery dates of the six leptons.

In 1927, thirty years after Thomson's discovery, physicists knew just three basic particles: the electron, the proton, and the photon. Although they had, by this time, accepted the reality of Einstein's light corpuscle (the photon), most physicists still didn't think of it as a "true" particle like an electron. That elevation in its status required the development of the quantum theory of charged particles interacting with light (quantum electrodynamics) in the late 1920s and early 1930s. This theory demonstrated convincingly that the photon had as much right as the electron to be called a fundamental particle. As for the proton, no one yet knew that it had a finite size or was composed of other, more fundamental particles.

The proliferation of particles that has continued to the present day began in the 1930s with the discovery of the neutron, the positron (antielectron), and mesons (particles of mass intermediate between the masses of the electron and proton). As revealed in work conducted in the late 1940s, many different particles were bombarding Earth in cosmic radiation, some indeed of intermediate mass (the mesons) and some that were heavier than protons.* Moreover, the particles were of two kinds, some that interacted strongly with nuclei and one that did not. Cecil Powell

---

* The "primary" cosmic radiation incident on Earth from outer space is relatively simple. It consists overwhelmingly of protons, with some heavier nuclei and some

and his group in Bristol, England managed to sort out the weakly interacting one. They did this by observing tracks of particles in sensitive photographic emulsions, finding that a somewhat heavier meson, the pi meson (later the pion) decayed into a somewhat lighter meson, the mu meson (later the muon), which in turn decayed into an electron. The pion and muon, despite being not far apart in mass, are about as different as two particles can be. The pion interacts strongly; the muon does not. The pion has spin zero; the muon has spin $\frac{1}{2}$. The pion has both charged and neutral versions; the muon is charged only. And, most important of all, the pion is a composite particle made up of a quark and an antiquark; the muon is a fundamental particle.

Soon after its discovery, physicists recognized that the muon is a "heavy electron." It seemed to share all of the electron's properties except mass. (And, indeed, it is *much* more massive, some 207 times heavier than an electron.) Still, being seven times less massive than a proton, the muon could be lumped with the electron as a "lightweight." The two, along with their still hypothetical neutrinos, were lumped together with the name lepton, derived from a Greek word meaning "small" or "light." No one had predicted that such a particle as the muon should exist. It was a total surprise. Reportedly, the noted Columbia University physicist I. I. Rabi asked about the muon, "Who ordered that?" And it is said that Richard Feynman, whose Nobel Prize was awarded for work in quantum electrodynamics, wrote on a corner of his Caltech blackboard the nagging question, "Why does the muon weigh [that is, why does it have so much mass]?" In truth, no one yet knows the answer to Feynman's question.

What's worse, the muon was not the end of "heavy electrons." When Martin Perl of Stanford University went looking for a still heavier charged lepton in the late 1960s, no one gave good odds that he would succeed. But he did, after a decade of painstaking work at the Stanford Linear Accelerator Center (now renamed the SLAC National Accelerator Laboratory). I won't go into the details of his subtle experiments here.

---

electrons. Many other particles are created in collisions high in the atmosphere, some of which reach Earth's surface to be studied in earthbound laboratories.

Martin Perl (b. 1927). In the early 1970s, when this photo was taken, Perl was not only embarked on the work that would earn him a Nobel Prize, he was also prodding the American Physical Society toward extending its reach to encompass issues of science and society. He was a founder of that organization's Forum on Physics and Society and was the first editor of its newsletter. In his eighties, Perl continues active research in physics. Photo courtesy of AIP Emilio Segrè Visual Archives.

They involved the creation of tau-antitau pairs when electrons and positrons collided. The total mass energy of an electron and a positron is about 1 MeV. In these SLAC collisions, the total kinetic energy of the colliding particles was 5 GeV, five thousand times greater than the mass energy. This kinetic energy is available to be transformed into the mass energy of new particles. As Table A.1 shows, more than 3.5 GeV is needed to create a tau and its equally massive antiparticle. An accelerator much less powerful than SLAC could not have done the job.

The electron, muon, and tau have come to be called three *flavors* of lepton. Although widely different in mass, they share spin ½ as well as the same weak interaction. To date, they appear to be fundamental (not composed of smaller entities) and without physical extension (that is, point particles). The three lepton flavors are also called *generations*, perhaps a more descriptive term and one that better aligns with the three generations of quarks.

I will save for Question 42 a discussion of the reasons that physicists are now pretty sure that three generations are all there are. When there were only two lepton generations known (electron and muon), no one had any idea whether there might be a third or an endless succession of other generations. Now, remarkably, there is good reason to think that three is it.

40. How many distinct neutrinos are there? How do we know? Once the muon was identified, physicists knew that it needed an accompanying neutrino. Like most particles, the muon is radioactive. It decays with a mean life of 2 microseconds (a good long time in the particle world) into an electron, a neutrino, and an antineutrino, a reaction that can be written

$$\mu^- \rightarrow e^- + \bar{\nu}_e + \nu_\mu.$$

The negative muon (symbol μ^-) decays into a negative electron (e^-), an antineutrino (symbol $\bar{\nu}$) and a neutrino (symbol ν). I put subscripts on the neutrino symbols to show that one of them is of the electron type and one is of the muon type. When muons were first studied in the late 1940s, however, there was no firm evidence for even one kind of neutrino, much less two. In their landmark experiment in 1956, Frederick Reines and Clyde Cowan, Jr. established the reality of the electron neutrino. For this they needed a powerful nuclear reactor to serve as a rich source of neutrinos (actually antineutrinos).

Let me trace out an example to show why a reactor produces a flood of antineutrinos. When a U-235 nucleus absorbs a neutron, it becomes a U-236 nucleus with some extra excitation. One of many possible results when this U-236 nucleus comes "unglued" is a nucleus of barium, $_{56}\text{Ba}_{86}{}^{142}$ (fifty-six protons and eighty-six neutrons), a nucleus of krypton, $_{36}\text{Kr}_{55}{}^{91}$ (thirty-six protons and fifty-five neutrons), and three free neutrons (which help sustain the chain reaction). These fission products are "neutron rich"; they contain more neutrons than do stable nuclei in this part of the periodic table. The barium nucleus undergoes beta decay, transforming itself into a nucleus of lanthanum (atomic number 57), which in turn undergoes beta decay, transforming itself into a stable nucleus of cerium, $_{58}\text{Ce}_{84}{}^{142}$. After these two radioactive steps, the number of protons has been increased by two, the number of neutrons decreased by two, and two antineutrinos have been emitted. The other fission product in this example, Kr-91, launches a chain of four successive beta decays, ending at a stable nucleus of zirconium, $_{40}\text{Zr}_{51}{}^{91}$. Another four antineutrinos have been emitted. In all, the fission of a single uranium nucleus has resulted in the creation of six antineutrinos, which come shooting out of the reactor, essentially undeterred.

This still doesn't answer the question of why vast numbers of antineutrinos are needed if even one is to be detected. It is because neutrinos are "all weak." The four kinds of interaction in the particle world (indeed in all the world)—strong, electromagnetic, weak, and gravitational—have quite different strengths. Some particles, such as protons, experience all of these interactions. So a proton interacts very readily with many other particles (apart from whatever inhibition electric repulsion between the particles might provide). By contrast, the electron, muon, and tau—the charged leptons—are immune to the strong interaction. They feel the other three. So an electron in an atom is influenced by the electric charge of the atomic nucleus but not by any strong force in the nucleus. When a charged lepton collides with another particle in an accelerator, it is primarily the electromagnetic interaction that comes into play. That is still sufficient to make for vigorous interaction. In the Stanford Linear Accelerator, for instance, electrons collide with positrons, making vast numbers of other particles.

Unlike the particles just mentioned, neutrinos feel only the weak force and the even still much weaker gravitational force. So neutrinos interact infrequently with other particles. It's all a matter of probability. Most low-energy neutrinos can penetrate the entire Earth (indeed, even light-years of solid matter) without deflection or interaction. The occasional one will travel a lesser distance before it interacts—miles or meters, or, with some extremely tiny probability, only centimeters. It's like a new car that might break down a block from the show room, although it is more likely to run flawlessly for thousands of miles. In their first definitive results, Reines and Cowan detected about three neutrinos per hour from among the billions per second coursing through their apparatus.

Since then, neutrinos have been detected from other sources, such as the Sun and high-energy accelerators—even, in 1987, from a supernova (an exploding star) in a nearby galaxy.

As work on muons went forward in the 1940s and 1950s, physicists had no doubt that there must be neutrinos as "fellow travelers" of both muons and electrons. There was some reason to believe that these neutrinos might be distinct, but there was no sure evidence that a single

Leon Lederman (b. 1922), shown here using an Einstein doll as a prop. Following his work with the team that discovered the muon neutrino, Lederman became director of Fermilab and also a tireless worker for improving science education in secondary schools. He helped propel the "physics first" movement (physics for ninth graders) now gaining currency in the United States.

Photo courtesy of AIP Emilio Segrè Visual Archives, W. F. Meggers Gallery of Nobel Laureates.

neutrino did not do double duty for both the electron and the muon. The clear evidence that the muon neutrino and electron neutrino are distinct particles came in 1962. Here is what the team headed by Leon Lederman, Melvin Schwartz, and Jack Steinberger did, in cookbook recipe form:

Preheat oven (Brookhaven Synchrotron) to an energy of 15 GeV.

Direct the proton beam from the synchrotron against a target to produce pions.

Let the pions run a short distance so that most of them decay into muons and neutrinos.

Erect a barrier that will stop the muons and other charged particles but let the neutrinos through (44 feet of iron and concrete should suffice).

Let the neutrinos that get through collide with atomic nuclei and see what happens.

Continue running for three hundred hours. Be patient.

In those three hundred hours, the experimenters found twenty-nine muons and no electrons created in the neutrino-nucleus collisions. If the muon neutrino were the same as the electron neutrino, it should have created about as many electrons as muons. It created only muons, showing itself to be distinct.*

When Martin Perl found the tau lepton in 1978, physicists, without exception, assumed that it, too, would have its own neutrino. In 2000, a team at Fermilab near Chicago verified the distinctness of the tau neutrino. Their method was not unlike that employed nearly forty years earlier at Brookhaven to pin down the muon neutrino, but they needed still greater energy because of the large mass of the tau lepton. Some tau neutrinos that were generated following the collision of 800-GeV protons with nuclei made it through thick barriers and magnetic fields to create a few tau leptons, whose tracks could be identified in photographic emulsions. No one was surprised, but everyone breathed a sigh of relief. There are three flavors of neutrinos, matching the three flavors of charged leptons.

41. Do neutrinos have mass? Why do they "oscillate"? When Wolfgang Pauli postulated the neutrino (or neutron, as he called it then) in 1930, he imagined that it might have a mass comparable to the electron's mass, but in any case much less than a proton's mass. Then came beta decay experiments that suggested that the mass of the neutrino might be zero. Consider, in particular, the decay of the triton—the nucleus of hydrogen-3. In reaction-equation form, it is

$$_1H_2^3 \rightarrow {}_2He_1^3 + e^- + \bar{v}_e.$$

With a half-life of about twelve years, the triton transforms into a helium-3 nucleus plus an electron plus an antineutrino. Like every spontaneous radioactive transformation, it is a downhill event in mass. The masses of the helium-3 nucleus, the electron, and the antineutrino (if it has any mass) must add up to less than the mass of the triton. Otherwise the decay would not take place. The difference between the initial mass and the sum of the final masses supplies the kinetic energy with which the final particles fly apart.

---

* For this discovery, Lederman, Schwartz, and Steinberger shared the 1988 Nobel Prize in Physics.

Because there are three final particles, the kinetic energy can be divided up in various ways. For instance, if the electron just dribbles out with very little kinetic energy, the neutrino will fly off with most of the kinetic energy. Conversely, if the neutrino dribbles out,* the electron takes most of the kinetic energy. Beginning in the 1930s and continuing up to the present day, physicists have measured the kinetic energy of the ejected electron with ever more precision—and, in particular, the maximum value of that kinetic energy. That maximum value of the *electron's* kinetic energy occurs when the *neutrino* has little or no kinetic energy, and, since the masses of the two nuclei and the electron are known with great accuracy, it tells us whether any of the available energy is used up in making the neutrino's mass. That mass would show up as a discrepancy when the energy books are balanced. All of the experiments, even the most recent ones, fail to find any energy taken up by the mass of the unseen neutrino. This led, initially, to the idea that the electron neutrino is truly massless. This didn't disturb anyone, since the massless photon was already a familiar friend. Moreover, some theoretical treatments worked better for a truly massless neutrino.

For better or worse, we are now sure that all three neutrinos have some mass. This mass doesn't show up in the triton decay experiment because it is so small. The present upper limit for the mass energy of the electron neutrino is 2 eV, about 250,000 times less than the mass of the electron. The upper limits for the individual masses of the muon neutrino and tau neutrino are much larger (see Table A.1), but the actual masses are probably much less than these limits. There is cosmological evidence that the total mass of all three neutrinos combined is less than 0.3 eV (which puts the average mass of each one at no more than one five-millionth the mass of an electron). The cosmological evidence comes from various astrophysical measurements. I won't go into the details, which are subtle. In one way or another, these bits of evidence are all related to the fact that neutrinos in the universe are so plentiful that if they have appreciable

---

Dribbles would not be a good word if the neutrino is strictly massless, for in that case it would always fly off at the speed of light, even if with very little energy.

mass, they will contribute substantially to the way the universe as a whole behaves: how its matter clumped into galaxies in the past, how evenly or unevenly its matter is now distributed, how gas clouds absorb radiation from distant galaxies.

Why are we so sure that neutrinos have mass if no measurement has provided a definitive value for the mass of any of the three of them, and the limits that we have on their masses are so exceedingly tiny? It's all about superposition. One of the features of quantum physics is that a particle or system can exist in more than one "state" at a time. The different states are said to be superposed. It's as if you could drive both northbound on one highway and eastbound on another highway at the same time. Actually quantum superposition is a bit more shadowy than that. It's more like your being in a dreamlike state, knowing that you wanted to go in a generally northeasterly direction, but not knowing whether you started out northbound or eastbound until a policeman with a radar gun stops you and asks why you are speeding northbound. Here's a simple example of superposition from the quantum world. An electron is emitted with its spin directed upward. It is in a spin-up state. But that spin-up state is a superposition of spin-left and spin-right states (see Figure 23). So if you later make a measurement to find out if the electron's spin is directed left or right, you will find (to your astonishment, if you happen to think classically) that indeed it *is* directed left or right. Half of such measurements will show it in a spin-left state and half will show it in a spin-right state.

Now to the neutrinos. It turns out that a neutrino in a definite flavor-state—that is, a neutrino that is definitely of the electron, muon, or tau type—is a mixture of three different mass states. So an electron neutrino does not have a definite mass; it is a superposed mixture of different (*slightly* dif-

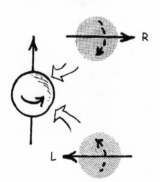

FIGURE 23
An electron with its spin up is an equal mixture of spin-left and spin-right states.

ferent) masses.* A neutrino made in the laboratory, or in the Sun, is in fact made in a definite flavor state and so in a mixture of mass states. The Sun, for example, emits only electron neutrinos. And a pion, when it decays into a muon, emits only muon neutrinos. Moreover, some modes of detecting neutrinos involve processes of definite flavor. Ray Davis, for example, in his pioneering work deep underground in the Homestake Mine in South Dakota, detected solar neutrinos by having them trigger the exact inverse of a beta decay process. This means that he was detecting only electron neutrinos.

Now to the last link in the chain of reasoning. A state of particular mass has a particular energy and therefore a particular frequency of vibration (in keeping with the Planck-Einstein formula $E = hf$, which applies to all systems, not just photons). Normally, for material systems, that frequency is too rapid to have a measurable effect. But, through a phenomenon very much like beat notes in music, it has a measurable effect if two or three different mass states are mixed. In the case of music, if two notes of slightly different frequency are sounded, a "beat note" is produced. The ear hears a tone whose frequency is the difference of the two original frequencies. Thus if the string of one violin vibrates at 440 Hz and the string of another violin vibrates at 442 Hz, you, the listener, hear a pulsating intensity of only 2 Hz. And so it is with neutrinos. An electron neutrino produced in the Sun, for example, is a superposition of three different mass states vibrating at three different frequencies. These frequencies "beat" to produce a pulsation manifested by that electron neutrino changing periodically into a muon neutrino, a tau neutrino, and back to an electron neutrino. That neutrino pulsation is called *neutrino oscillation*. The measured fact of neutrino oscillation is taken as overwhelming evidence that neutrinos have mass and that the three mass states are not equal. If neutrinos were truly massless, or if they all had the same mass, oscillation would not occur.

---

*As you might guess, conversely, a neutrino of definite mass, if we could make one, would be a superposition of neutrinos in three different flavors.

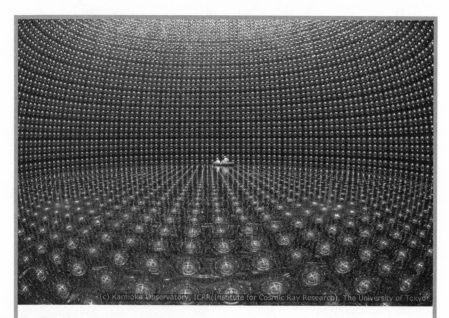

The Super Kamiokande detector with its more than eleven thousand photodetectors, partly filled with pure water. Photo courtesy of Kamioka Observatory, ICRR (Institute for Cosmic Ray Research), University of Tokyo.

The fact that Ray Davis found only about one-third as many electron neutrinos as he expected from the Sun didn't prove that neutrinos oscillate, but it was suggestive. (Initially no one could explain the deficit, and the astrophysicist John Bahcall is said to have remarked, "If Ray Davis finds any fewer neutrinos, he will have proved that the Sun doesn't shine.") The first strong evidence for neutrino oscillation came in 1998 from Super Kamiokande, a deep underground detector in Japan. Researchers there measured muons produced by muon neutrinos, which themselves were produced high in the atmosphere in cosmic-ray collisions. Since Earth provides no impediment at all to neutrinos, the Kamiokande detector should have been bathed in neutrinos from all directions, underfoot as well as overhead, and from all sides. Instead the researchers found fewer muon neutrinos arriving from the far side of Earth than from the near side. Those traveling 8,000 miles were being depleted more than those traveling only a hundred miles or

Ten thousand photodetectors on this SNO ball peer inward at the 1,100 tons of heavy water it contains, more than a mile underground. Photo courtesy of Ernest Orlando Lawrence Berkeley National Laboratory.

so. This was evidence that during that longer trip, some muon neutrinos were oscillating into something else, thus escaping detection.

Then, in 2001 and thereafter, a more sophisticated detector—SNO, or the Sudbury Neutrino Observatory in Sudbury, Ontario, Canada—found even more convincing evidence for neutrino oscillation.* In SNO, neutrinos from the Sun bombard a tank containing 1 million kg of heavy water. That's a lot of deuterons—nuclei of heavy hydrogen consisting of one proton and one neutron. If a solar neutrino causes the proton within a deuteron to turn into a neutron and a positron, that is definitive evidence for the neutrino being of the electron flavor. If the neutrino merely "bounces" from the deuteron (or scatters, as physicists say), it could be of any flavor. The SNO results show that the total number of arriving neutrinos from the Sun is three times the number of the electron flavor. Oscillation (which, incidentally, has now been checked in other ways as well) is surely real.

And if there is oscillation, there is mass—indeed at least two, and more likely three, different masses. Physicists like to ponder why things are as they are. The best guess is that the mass of neutrinos arises from a yet-to-be-discovered particle called the *Higgs boson* (Question 98).

42. Are there really only three generations of particles? First, let me say that quarks, like leptons, line up in three generations. And just as there are two leptons in each generation, there are also two quarks in each generation.† Physicists, who like order and simplicity, are happy about this (even if neutrino mass and neutrino oscillation hardly seem very simple). But how do we know—or, better, why do we believe—that there is no fourth generation or a whole series of additional generations? After all, the second generation of leptons (the muon and its neutrino)

---

* SNO is located in a nickel mine. Other, earlier neutrino experiments were conducted in a gold mine in South Africa as well as in the Homestake gold mine in South Dakota. The mining industry has been good for neutrino physics.
† Apparently for the sole purpose of keeping physicists on their toes, however, nature has arranged for the six quarks to have six different flavors, whereas three flavors suffice to characterize the leptons.

came as a surprise, and the third generation (the tau and *its* neutrino) as an even bigger surprise. The quark story is similar: Two quarks were enough to account for protons and neutrons, and a third quark was needed to explain the so-called strange particles. These three were followed by a fourth quark, the charm quark (completing generation 2), and finally by the heavier and rarer top and bottom quarks (generation 3).

Surely, with these additional generations sneaking up unbidden, should not physicists anticipate yet more generations to come, just as an anthropologist encountering the remains of one and then a second and then a third temple in the jungle might reasonably expect that still more await discovery? Several things put the lid on that anticipation. One is the fact that the SNO experiments (consistent with early indications from the Ray Davis experiment) find that one-third of the neutrinos hurtling through their apparatus from the Sun are of the electron flavor. This suggests a total of three flavors (or generations). If there were ten flavors, one would expect oscillation among them to leave only one in ten having the electron flavor after their 150-million-kilometer journey from the Sun.

The most convincing evidence for just three generations is more subtle but remarkably compelling. An uncharged particle called the Z^0 (*zee-naught*), which is nearly one hundred times more massive than a proton, decays into various pairs of particles. Some of these pairs, such as an electron and a positron (antielectron) or a muon and an antimuon, can be seen in the laboratory because they are charged and interact readily in a detector. Other pairs, such as an electron neutrino and its antiparticle, go undetected because they are neutral and interact only very weakly with matter. Experimenters can measure the probabilities of the various decays into charged particle pairs, and they can also measure (I will explain how in a moment) the grand total probability of decay for the Z^0 particle. From the difference, they can infer the probability of decay into pairs of neutral particles.

One of the fascinating features of quantum physics is captured in the Heisenberg uncertainty principle* (advanced by the twenty-five-year old

---

*See Question 74.

Werner Heisenberg in 1927). In one of its forms it says that time and energy cannot be simultaneously measured to arbitrary precision. The more accurately the time of an event is known, the less accurately its energy is known. The more accurate the energy, the less accurate the time. There is a direct implication for short-lived particles such as the Z^0, whose lifetime is only about 10^{-25} seconds (in that time, light would not even make it from one side of an atomic nucleus to the other). The particle's energy—which really means its mass—is rendered measurably uncertain because of the extremely short time it lives. When physicists measure the mass of the Z^0 repeatedly, they find a range of values spanning some 2.5 GeV (the average mass is 91 GeV). Using the uncertainty principle, they infer the lifetime, and from the lifetime the total probability of decay into all possible products. The amount by which this total probability exceeds the measured probability of decay into charged-particle products tells the probability of decay into unseen neutral particles.

Analysis of the Z^0 data leads to the quite astonishing conclusion that the number of flavors of uncharged leptons (that is, neutrinos) is 2.92, with an uncertainty around 2 percent. There is only one way to interpret this result: that the number of lepton flavors (or generations) is 3.

43. How do we know that all electrons are identical? You might say, "Well, an electron is an electron. Of course they are all the same. If two particles were not the same, physicists would not call them by the same name." That is not an adequate answer, because the word *identical* in the question means the same in every particular, *absolutely* the same. That makes the question deep and important, because it relates to how many additional layers, if any, remain to be discovered in nature.

In the ordinary world around us, *no* two things are ever identical. To satisfy the National Basketball Association and its teams, a manufacturer of basketballs tries to make identical basketballs. They are shipped from the factory all with the same diameter and the same mass, made of the same materials and ready to be inflated to the same pressure. But, of course, they are far from identical. Examined at the atomic level, every one differs in billions of particulars from every other one. From a particle physicist's point of view, the "identical" basketballs are a kaleidoscope of variety. The same is true of "identical" twins. Many of their strands of DNA are indeed identical, but the twins themselves differ in myriad ways.

Yet when we probe into the particle world, we seem to find true identity. Think first how many quantities are needed to describe an electron fully: its mass, its charge, its spin, its flavor, its magnetic strength, the strength of its weak interaction. That's about it. The simplest thing you could name in your own environment would require vastly more quantities for its full description (down to the atomic level). So as entities get smaller, they seem to get simpler. This is not just an obvious truism. If there were infinitely many layers to uncover, the electron could be about as complex as a basketball or the whole Earth. Moreover, these few quantities that characterize the electron are known with stunning precision—to better than one part in a million for the charge, mass, and magnetic moment. This means that every batch of electrons studied by one researcher matches exactly the properties of every other batch studied by every other researcher.

The most compelling evidence for the identity of all electrons is that they obey the Pauli exclusion principle. The usual statement of this principle is that no two electrons can occupy the same state of motion at the same time. More precisely, the principle is this: No two identical particles of half-odd-integral spin (this means no two identical fermions) can occupy the same state of motion at the same time. The principle applies to *identical* particles only (provided they are fermions). The fact that this principle is obeyed by electrons, with no known exceptions, is strong evidence that electrons are identical.

So, at the deepest level in particle physics, we find entities that are truly identical and that are fully described by relatively few properties. This is suggestive of a floor, a bottom, to our exploration. Whether we are near that floor now or will find it at the far smaller dimensions being studied by string theorists, no one can say, but it seems likely that there really is a floor.

section VIII

And More Particles

44. Names, names, names. What do they all mean? One of the impediments to learning about particles is nomenclature. There is a confusing array of names, some of them whimsical. Here is a glossary of some of the terms used by particle physicists. It should help as a reference point. Some of these terms have appeared in earlier sections; some appear in this and later sections.

> *Leptons*—Fundamental particles of spin $\frac{1}{2}$ that do not feel the strong interaction. There are six leptons: three of charge −1 (electron, muon, and tau), and three that carry no charge (the electron neutrino, muon neutrino, and tau neutrino).
>
> *Quarks*— Fundamental particles of spin $\frac{1}{2}$ that do feel the strong interaction. There are six quarks: up, down, charm, strange, top, and bottom. The up, charm, and top quarks have charge $+\frac{2}{3}$. The down, strange, and bottom quarks have charge $-\frac{1}{3}$. Quarks are never seen in isolation. They combine in twos or threes to make *hadrons*.
>
> *Force carriers*—Fundamental particle of spin 1 that mediate interactions. There are eleven force carriers: the W (with positive and negative manifestations) and Z^0, which mediate the weak interaction; the photon, which mediates the electromagnetic interaction; and eight gluons, which mediate the strong interac-

tion. (There is also a theoretical graviton, of spin 2, that mediates the gravitational interaction but—so far—plays no role in the particle world.) Force carriers are also called *exchange particles*.

Flavor—An arbitrary name used to sort out different kinds of *leptons* and *quarks*. Leptons have three flavors. Each charged lepton and its associated neutrino constitute one flavor. All six quarks are assigned different flavors. (The words *family* and *generation* have also been used to segregate the different kinds of fundamental *fermions*.)

Color—An arbitrary name used to sort out different forms in which each *quark* can manifest itself. There are three colors, conventionally called *red, green,* and *blue.*

Hadrons—Composite particles made of two or more *quarks*. Unlike quarks, they can be observed singly in the laboratory. They interact strongly.

Mesons—*Hadrons* of integral spin made of a *quark* and an antiquark. Mesons are *bosons.*

Baryons—*Hadrons* of half-odd-integral spin (for example, $\frac{1}{2}$, $\frac{3}{2}$, $\frac{5}{2}$) made of three *quarks*. Baryons are *fermions.*

Pion—The least massive *meson.*

Nucleon—Either a proton or a neutron, the two least massive *baryons.*

Fermion—Any particle of half-odd-integral spin (for example, $\frac{1}{2}$, $\frac{3}{2}$, $\frac{5}{2}$). Fermions obey the Pauli exclusion principle.

Boson—Any particle of integral spin (for example, 0, 1, 2). Bosons do not obey the Pauli exclusion principle.

Antiparticle—An entity identical to a particle in some properties (for example, mass) and opposite in other properties (for example, charge). The antiparticle of an antiparticle is the original particle. Most particles have a distinct antiparticle. For a few (notably the photon), particle and antiparticle are identical.

45. What are the properties of quarks? How do they combine?

The decades right after World War II saw a rapid increase in the number of denizens in the particle zoo. Among these were "strange" particles, so named because they were created rapidly in particle collisions—evidently by strong interactions—but they decayed slowly as if the decay were governed by weak interactions. One strange particle, for example, is the lambda-naught, whose mass of 1,116 MeV is enough to allow it to decay into a proton (mass 938 MeV) and a pion (mass 140 MeV):

$$\Lambda^0 \rightarrow p + \pi^-.$$

The mean life of this decay, 2.6×10^{-10} second (less than a billionth of a second), qualifies the decay as "slow." If the decay were in the grip of the strong interactions and unrestrained, the half-life would be expected to be billions of times shorter.

The oddly slow decay of the strange particles was one of the factors that led Murray Gell-Mann and George Zweig to suggest independently* in 1964 that hadrons were composed of smaller constituents, which Gell-Mann called *quarks*. They postulated three quarks, called *up*, *down*, and *strange* (*up* and *down* having no relevance to directions in space), and realized that these new particles, if they were real, must have fractional charge. The charge of the up quark is $+\frac{2}{3}$, that of the down and strange quark $-\frac{1}{3}$. In their scheme, a baryon—such as a proton, neutron, or lambda—contains three quarks, and a meson—such as a pion—contains a quark and an antiquark. Using the shorthand u for up, d for down, and s for strange, the compositions postulated by Gell-Mann and Zweig (and now well-established) are uds for the lambda, uud for the proton, and d$\bar{\text{u}}$ for the negative pion. The bar over the u indicates the antiquark. Then the reaction equation for the lambda decay can be written

$$\text{uds} \rightarrow \text{uud} + \text{d}\bar{\text{u}}.$$

By reference to Table A.2, it is easy to see that the charges combine correctly: zero for the lambda, +1 for the proton, and −1 for the negative pion (keeping in mind that the charge of the antiquark $\bar{\text{u}}$ is $-\frac{2}{3}$, opposite to the charge of the quark u). The reaction equation also reveals a new

---

*Both were at Caltech at the time and both were thinking about the same puzzles in particle physics, but they arrived separately at the idea of quarks.

Murray Gell-Mann (b. 1929), to whom we owe the name *quark*. As a child prodigy, Gell-Mann's interests lay in linguistics and archaeology, fields he pursued alongside physics throughout his life. He earned his bachelor's degree from Yale at eighteen and his Ph.D. from Massachusetts Institute of Technology at twenty-one, becoming a full professor at Caltech by age thirty. In his later years, he recentered his physics research in the field of complexity. Photo courtesy of AIP Emilio Segrè Visual Archives, *Physics Today* Collection.

rule: When physicists keep track of numbers of particles, they have to count antiparticles as negative. On the left side of the reaction equation are the three quarks that compose the lambda particle. On the right side, after the decay, there are four quarks and an antiquark. The number of quarks is preserved if the antiquark is assigned a negative particle number. Here is another example illustrating the same idea. When an electron and a positron collide and mutually annihilate to create two photons, the lepton number after the collision is clearly zero, for there are no leptons left. And the lepton number is also zero before the collision if the electron is counted as +1 and its antiparticle, the positron, is counted as −1. Such a rule seems arbitrary, but it works in all cases where particle numbers are preserved, and it has a theoretical basis.

If, in the above reaction equation, you examine not just the total number of quarks before and after, but the numbers of specific kinds of quark, you see something interesting. The number of up quarks is the same before and after (+1 before and $+2 - 1 = +1$ after), whereas the number of down quarks increases by 1 (from 1 to 2) and the number of strange quarks decreases by 1 (from 1 to zero). It appears that in the decay a strange quark has changed into a down quark. Indeed this is a good description of what is happening, and it is the change of quark type that puts the brakes on the decay and stretches it out to almost a billionth of a second. Strangeness is a property that is "partially conserved." The strong interaction leaves it unchanged, but it is malleable in the hands of the weak interaction. (More at Question 60.)

The three quarks I have just been discussing are only half of those that are now known. I will skip the historical details of how the other three were predicted and then found. As to dates, the charm quark made its appearance in 1974, the bottom quark in 1977, and the top quark in 1995. The deepening understanding of quarks, although not exactly in lockstep with lepton discoveries, paralleled those discoveries. First the muon joined the electron and its neutrino to make three leptons, just as the strange quark joined the up and down quarks to make three. The muon's neutrino brought the number of leptons to four, as the charm quark brought the number of quarks to four. And for both leptons and quarks, a final pair (the tau and its neutrino for leptons, the bottom and top quarks for the quarks) rounded out the picture to a half-dozen of each, a number that we now have reason to think is probably the end of the line.

As I have mentioned before, the six quarks, with their six flavors, are divided into three generations, matching the three lepton generations. Down and up constitute the first generation, strange and charm the second, bottom and top the third. Their very disparate masses (see Table A.2) reflect this generational hierarchy. The leptons are similarly divided into three sets of two. However, for leptons the generational division is really the same as the flavor division.

Besides flavor, quarks have another quantum property called *color*. (Needless to say, a quark's color has nothing to do with actual colors in our large-scale world.) All six of the quarks can be in one of three states called *red*, *green*, and *blue*. Observed entities—such as protons, pions, neutrons, or lambdas—are "colorless." *Colorless* really means an equal mixture of the three colors, so at any one instant the three quarks making up a proton are in three different color states. One is red, one green, and one blue. What's more, antiquarks can be antired, antigreen, and antiblue.

The fact that quarks are colored means that the gluons that hold them together are also colored, but in a peculiar way. Each gluon has one color and one anticolor, so red-antigreen and blue-antired are possible descriptions of a gluon. When a gluon is emitted or absorbed by a quark, the quark may change color, but always in a way that keeps the "large" composite particle colorless. The coloration of gluons accounts for their number—eight in all (even though simple figuring with three colors and three anticolors might suggest nine).

Neutrinos are often said to be elusive because of the weakness of their interaction and their ability to penetrate miles of solid material as if it were empty space. In a way, quarks are even more elusive. They interact strongly, to be sure, but they remain securely hidden within the particles we do see. No single quark has ever been seen in the laboratory. I discussed the reason for this at Question 9.

46. What are the composite particles? How many are there?

When two or more fundamental particles join in close proximity to make another particle, we call the result a *composite* particle. Since leptons experience no strong force that would bind them together in a small space,* all of the composites are made of quarks (including antiquarks). (Will there be deeper layers? Will quarks someday show themselves to be composite? It doesn't seem likely, but who knows?)

Mesons are made of quark-antiquark pairs. Some are shown in Table A.4 in Appendix A. Baryons are made of three quarks.† A few of them are also shown in the table. Since there are only so many ways that a half-dozen different quarks can be arranged in twos and threes, you might think that there must be a limited (even if rather large) number of composite particles. Nature thwarts that expectation. Quark combinations, like atoms, can exist in ground states and excited states. Numerous excited

---

*The nearest we come to a composite particle made of leptons is *positronium*, the name given to an electron and a positron bound together by their electric attraction. This entity, occupying even more space than a hydrogen atom, is understandably called a positronium *atom*, not a positronium *particle*.
† Since 2003, various researchers have reported evidence for a "pentaquark," a baryon composed of four quarks and an antiquark, but recent searches failed to find it. If it exists, it has an extremely short half-life and is exceedingly hard to pin down.

states of the nucleon, for example, are known. These decay back to a nucleon not by emitting a photon, as an excited atom does, but, typically, by emitting a pion. The lifetimes of these excited composites are extremely short. Nevertheless, their number is legion. So the answer to the question "How many are there?" is "A limitless number."

The omega-minus (Ω^-) is an interesting baryon because it is "triply strange," being composed of three strange quarks. Its existence was predicted by Murray Gell-Mann (even before he proposed quarks) and its reality was confirmed in 1964 by a group headed by Nicholas Samios of Brookhaven National Laboratory. By the time of the discovery, physicists believed there were three quarks, and it nicely rounded out the set of particles that could be made from up, down, and strange quarks. The Ω^- also provides supporting evidence for color as a quark property. If the three strange quarks composing the Ω^- were identical, the Pauli exclusion principle would prevent them from occupying the same state of motion. If they combined, one of them would have to have greater energy, like the third electron in the lithium atom (the first two can share the same energy state by having their spins oppositely directed). The relatively low energy (low mass) of the Ω^- suggests that the three quarks composing it may in fact be in the same state of motion. This is possible if they differ in the quantum property called *color*. Since the Ω^-, like all observed particles, is "colorless," the three quarks that compose it must, at any moment, be red, green, and blue.

The predominant mode of decay for the Ω^-, as shown in Table A.4, is into a lambda-naught and a negative kaon:

$$\Omega^- \rightarrow \Lambda^0 + K^-.$$

In terms of quark constituents, this reads

$$sss \rightarrow uds + \bar{u}s.$$

You see that the number of strange quarks has decreased from three to two, and the number of down quarks increased from zero to one. So, just as with the decay of the Λ^0, a strange quark has transformed into a down quark, a change of strangeness that explains why the Ω^- lives so long.

(Note that in this decay, the number of up quarks remains zero, since after the decay there is one up quark and one anti-up quark.)

The kaon resulting from the decay of the Ω^- is an example of a strange meson. It consists of a strange quark and an anti-up quark. Its decay is also slow because of the change of strangeness, which is driven only by the weak interaction. The charged kaon typically decays into a muon and an antineutrino of the muon flavor:

$$K^- \rightarrow \mu^- + \overline{\nu}_\mu.$$

In terms of quarks, the decay is

$$\overline{u}s \rightarrow \text{no quarks.}$$

The quark number is zero before and after. Indeed, *all* mesons have quark number zero, and all are unstable.

The sizes of composite particles may vary, but most are probably roughly the same size as the proton and neutron—that is, about 10^{-15} m. This distance, about one hundred thousand times smaller than the size of a typical atom, seems small. It *is* small. Yet it is large enough to hold an assemblage of quite a few quarks and gluons and be home to vigorous activity.

Composite particles are created in profusion in high-energy accelerators. The energy leader for many years was the Tevatron at Fermilab near Chicago. There protons and antiprotons, each with an energy of 1 TeV, collided head-on, making 2 TeV, or 2,000 GeV, or 2 million MeV available to create new particles. Not surprisingly, dozens or even hundreds of particles can be created in a single proton-antiproton collision at this energy. The Large Hadron Collider at CERN in Geneva, Switzerland is donning the mantle of most powerful as it ramps up to full power after 2012. By the time you read this, it may be directing 7-TeV protons against other 7-TeV protons for a total available energy seven times greater than that at Fermilab.*

---

*The Superconducting Super Collider was, regrettably, killed by the U. S. Congress in 1993. In a tunnel with a circumference of fifty-four miles near Waxahachie, Texas, its protons were scheduled to reach an energy of 20 TeV, providing 40 TeV when they collided.

The four-mile-long ring of Fermilab's Tevatron and its smaller injector ring (foreground) are buried at a site west of Chicago. Photo courtesy of Fermilab.

Two accelerators, one in California and one in Japan, are set up to function as "B factories," meaning to make large numbers of B mesons, whose relatively large mass of more than 5 GeV turns out to make them well-suited for studies of subtle features of the weak interaction. A B meson is composed of an antibottom quark and a first- or second-generation quark (up, down, strange, or charm). It has myriad decay modes. In one of these modes, the B meson's antibottom quark turns into an anticharm quark. In other modes, a lepton-antilepton pair is emitted. In every case, the decay is guided by the weak interaction, making the B's half-life a lengthy 10^{-12} second. The particular feature of the weak interaction elucidated by the B decay is the violation of what is called CP *conservation* (Question 61).

47. Does every particle have to be a fermion or a boson? What sets these two classes apart? Indeed every particle (at least everything that a physicist calls a particle) is a fermion or a boson. But not every *entity*

Part of the Large Hadron Collider's 17-mile-long tunnel beneath Switzerland and France. The large cylinder holds the magnets that steer the beam. In the foreground is a cutaway view showing an artist's rendering of the side-by-side beam tubes and proton beams (the thin white lines). Photo © CERN.

is a fermion or a boson. A basketball is neither, an automobile rolling off the assembly line is neither, a grain of sand is neither. It all has to do with identity. If two or more things are identical—truly identical in every respect—those "things," we have reason to believe, must be either fermions or bosons. No two basketballs, automobiles, or grains of sand are truly identical. They escape the fermion-boson classification. On the other hand, every electron is identical to every other one (or so we believe, and there is plenty of evidence that this is true). Electrons are fermions. And every pion gives evidence of being identical to every other pion. Pions are bosons.

To be a fermion or boson, a "thing" doesn't have to be as small as an electron or a pion. It could be as huge as an atom or molecule. All that is important is that there be identical siblings of the "thing." An atom of rubidium-85, for example, is a boson. There is a rule for figuring out whether an atom is a boson or a fermion. Count the number of fermions of which it is composed. If that number is even, the atom is a boson. If the number is odd, the atom is a fermion. Note that the basic constituents of every atom and molecule are fermions—electrons, protons, and neutrons, or, if you go down a layer, electrons and quarks. In either case, the odd-even arithmetic comes out the same. Rubidium is element number 37, so an atom of rubidium-85 contains 37 protons, 48 neutron, and 37 electrons—a total of 122 fermions, making the combined entity a boson. (Alternatively, you can count up 255 quarks and 37 electrons in the atom for a total of 292 fermions, again an even number.) By contrast, an atom of uranium-235, with 235 nucleons in its nucleus and a surrounding cloud of 92 electrons, is made of an odd number of fermions and is itself a fermion.

How is all this important, apart from providing "naming opportunities"? That's where quantum physics comes in. Imagine a tiny box containing two particles and nothing else. Particle 1, let's say, is in state of motion A, and particle 2 is in state of motion B. If the particles are distinct, that's about all there is to say about their combined motion. But what if they are identical? Then you don't know if particle 1 is in state A and particle 2 in state B or if particle 2 is in state A and particle 1 in state B. In fact, you *can't* know. Because the particles are identical, there is no way to specify for sure which particle is in which state. The quantum physics answer is that *both* particles are in *both* states. They are superposed. Think, for instance, of a helium atom in which one electron is in the ground state and one electron is in an excited state. You know that there is one electron in each state, but you don't know—and *can't* know—which electron is in which state.

In quantum theory, the state of a particle or a system is represented by a "wave function." In Classical physics, every quantity with physical meaning can be measured—quantities such as mass or speed or electric charge or magnetic field, and many others. Each such quantity is an "observable." Not so in quantum theory. The wave function is an example of a quantity that is important and meaningful but that is *not* an observable. Rather it is the *square* of the wave function that is observable.* That

---

*Technically, it is the absolute square of the wave function that is measurable. The wave function itself can be what mathematicians call *complex*—a number

squared quantity gives the probability of finding the particle in a particular state or at a particular place. This subtle distinction between classical physics and quantum physics is all-important to the behavior of bosons and fermions.

Now I have to indulge in just a little bit of mathematics and hope it clarifies more than it confuses. If particle 1 is in state A and particle 2 is in state B, the wave function representing that situation can be written, very schematically, as A(1)B(2). But, as I indicated above, because of the identity of the particles, switching 1 and 2 doesn't change anything. The wave function A(2)B(1) has to describe the same state of affairs. One way that quantum theorists deal with this ambiguity is to say that the wave function for the two identical particles in states A and B is the sum

$$A(1)B(2) + A(2)B(1).$$

You can see in this combination that if 1 and 2 are switched, nothing changes. But there is another possibility. The wave function for the combined state of identical particles in states A and B can also be written as the difference

$$A(1)B(2) - A(2)B(1).$$

Now, if you interchange 1 and 2, you get the negative of what you had before, but that is all right because the physically meaningful quantity, the observable, is the square of the wave function, and the squares of a quantity and its negative are the same.

These two possibilities, the sum and the difference, don't exhaust all the ways in which a combined wave function might be constructed, but it seems that nature has opted for just these two possibilities. Particles whose combined wave function is the sum of the two terms are called *bosons*. Particles whose combined wave function is the difference of the two terms are called *fermions*. So far as we know, all particles—all particles that have myriad identical copies of themselves—are either bosons or fermions.

---

with both real and imaginary parts. This is permissible, as the wave function is not itself a measurable quantity.

There is one last and important argument to be made in this mathematical diversion. What if states A and B are the same? This means that not only are the particles identical, they also occupy the same state of motion. For bosons, the above summed wave function would then read

$$A(1)A(2) + A(2)A(1).$$

No problem. That is the same as $2A(1)A(2)$. For fermions, on the other hand, the wave function with the negative sign would read

$$A(1)A(2) - A(2)A(1).$$

Big problem. This combination is zero. Two identical fermions *cannot* occupy the same state of motion. When Wolfgang Pauli first proposed the exclusion principle for electrons, he was not thinking along these lines. He was only trying to account for features of atomic structure. The explanation of the exclusion principle in terms of wave functions came later. In fact, some fifteen years after his initial proposal, Pauli himself provided a mathematical derivation showing that particles of half-odd-integral spin (such as electrons and protons) are fermions and obey the exclusion principle, whereas particles of integral spin (such as photons and pions) are bosons that do not obey the exclusion principle.

It is a sobering thought that something as simple as a minus sign in an *unobservable* wave function can account for the exclusionary behavior of electrons that is responsible for the entire structure of the periodic table and for the fact that you and I are here to think about it.

48. What is a Bose-Einstein condensate? The ninety-two electrons in a uranium atom can be compared to ninety-two residents of a high-rise apartment building, all of whom want to live as close as possible to street level (for the electrons, that means "wanting" to have as low an energy as possible). If the landlord imposes on the residents an "exclusion principle" that limits every apartment to at most one resident, the apartments will be occupied from the ground up to apartment number 92, and all apartments above number 92 will be empty. The apartments, of course, correspond to states of motion. If energy is added to "excite" the atom, an electron can jump temporarily to a higher-energy state. It's as if an exuberant apartment dweller, despite a preference for lower floors, moves temporarily to one of the unoccupied apartment farther up.

For electrons in a metal, the situation is similar, except that the number of electrons, even in a small bit of metal, is measured in the countless trillions and the spacing between energy levels is measured in nearly infinitesimal increments. Yet, despite the vastly larger number of electrons and the vastly smaller spacing between energy levels, the exclusion principle works its inexorable will in the same way. Electrons "pile up" in energy until the last one sits at the highest occupied level, with all states below occupied and all states above empty.

At least, that's the way it would be at or near the absolute zero of temperature—a sharp dividing line between occupied and unoccupied states. At finite temperature, the thermal energy of the electrons also comes into play. Then a few of the electrons near the top of the energy "pile" jiggle into higher states, leaving a few states unoccupied. It's as if, in the apartment house, there is a kind of "transition zone," a few floors that are partially occupied. Above this zone all apartments are empty. Below it all are fully occupied.

What if the apartment-house dwellers are bosons instead of fermions? Then the landlord imposes no exclusion principle. The residents can cluster in the ground-floor apartment. As it turns out, bosons not only *can* occupy the same state, they even have some preference for doing so. This suggests that indeed all would wind up together in the lowest state of motion. But again thermal motion comes into play. The bosons can get together and truly share a single state—a ground state—only if the energy available to each one is extremely small. Otherwise the thermal jiggling keeps many of the bosons in higher-energy states, more than offsetting their proclivity for clustering.

When a gas made up of bosons such as the rubidium-85 atoms I referred to at Question 47 is cooled to near absolute zero, many atoms can in fact share a lowest state of motion, clustering into a completely overlapping—or, perhaps better, interpenetrating—blob in which no atom

Carl Wieman (b. 1951) and Eric Cornell (b. 1961) in Boulder Colorado in 1996, with justifiable smiles following their success the previous year in creating a Bose-Einstein conden- sate in a gas of rubidium atoms. Wieman won not only the Nobel Prize for his achievements in physics research but also the Oersted Medal of the American Association of Physics Teachers for his contributions to physics education. In 2010 he joined the White House Office of Science and Technology Policy. Cornell, whose first exposure to particle physics as a Stanford undergraduate was not to his liking, now studies the behavior of ultracold fermions as well as bosons in his Boulder laboratory. Photo by Ken Abbott; courtesy of the University of Colorado, Boulder.

can be distinguished from any other. It's as if the apartment-house resi- dents are not only all sharing the ground-floor apartment but are being transformed into murky clouds, each one of which spreads throughout the apartment. (But the blinds had better be drawn. If the least bit of energy gets into the apartment, the residents will revert to their indi- vidual selves.) Such a clustering is called a *Bose-Einstein condensate*. Al- bert Einstein predicted this phenomenon in 1924, and it provided the explanation for the superfluidity of liquid helium (see Question 89), dis- covered in 1938. But many more years elapsed before researchers could realize a Bose-Einstein condensate in a gas of freely moving particles. In 1995 Carl Wieman and Eric Cornell, working in Boulder, Colorado, demonstrated the phenomenon in a gas of rubidium atoms. Not until then were the tools available to cool a gas of atoms (several thousand atoms, in Wieman and Cornell's first trials) to within a fraction of a mil- lionth of a degree of absolute zero. The photo on the next page shows data indicating the onset of Bose-Einstein condensation at a tempera- ture of 200 nK (200 billionths of a degree above absolute zero).

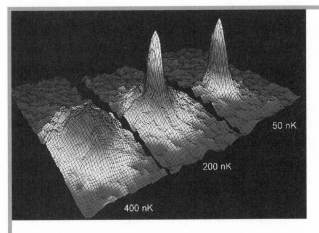

Data map by Wieman and Cornell showing the velocity distribution in a cloud of rubidium atoms as the temperature is lowered from 400 to 200 and then to 50 nK (billionths of a degree above absolute zero). The peak that appears when the temperature is 200 nK or lower indicates the formation of a Bose-Einstein condensate in which the atoms move at near-zero velocity. Image courtesy of Mike Matthews, Carl Wieman, and Eric Cornell, University of Colorado, Boulder.

Since then, many other researchers, starting with Wolfgang Ketterle*
at the Massachusetts Institute of Technology, who used sodium atoms
(sodium-23, with eleven protons, twelve neutrons, and eleven electrons,
is a boson), have achieved Bose-Einstein condensates. Will these condensates (or BECs, as they are often called) find practical application? It
would be dangerous to say no. More often than not, when scientists have
gained understanding of and control over some new form of matter, they
have found a way to harness it for some practical purpose. But in this
case it will be challenging.

49. How did bosons and fermions get their names? In 1924, Satyendra Nath Bose (BO-suh), a thirty-year-old physicist at the University of
Dacca,† sent a letter to Albert Einstein, who was then in Berlin, and at-

---

* The 2001 Nobel Prize in Physics was shared by Weiman, Cornell, and Ketterle.
† Dacca, now the capital of Bangladesh and written *Dhaka*, was then in India.

tached to it a paper titled "Planck's Law and the Hypothesis of Light Quanta." This paper, written in English, had just been rejected by a leading British journal, *Philosophical Magazine*. Who knows why? Because it came from a nobody in a distant land? Because it dealt with "hypothetical" light quanta (for which, by that time, the evidence was actually quite strong)? Because it derived an already-known result, albeit in a new way—Planck's law of the distribution of energy among different frequencies in a radiation-containing cavity? Bose, who *knew* that he had done something significant even if the British editors didn't know it, decided to approach the most famous living physicist (as it happened, Bose had translated Einstein's relativity text from German to English for the Indian market). Einstein read the paper, immediately appreciated its importance, personally translated it into German, and arranged for its prompt publication in a leading German journal, *Zeitschrift für Physik*.

What Bose had done was derive Planck's law by assuming that the radiation in a cavity consists of a "gas" of "light quanta" (that is, photons, in modern terminology) that did not interact with one another and could occupy any energy state, irrespective of whether another light quantum already occupied the state.* The particles in the gas had to conserve energy (when one quantum got more, another got less) and obey general principles of statistical mechanics.

Einstein recognized right away that Bose's work provided a third pillar of support for the light quantum, which Bose still called a hypothesis and which Einstein, who had proposed it nearly twenty years earlier, no doubt believed to be quite real. The first two pillars were the photoelectric effect, in which each light quantum absorbed at a metal surface gives all its energy to an electron, and the so-called Compton effect discovered by the American physicist Arthur Compton in the previous year, in which an X-ray photon bounces from an electron in an atom much as one billiard ball bounces from another.

---

* Bose had no inkling that an exclusion principle might apply to some kinds of particles. Pauli's postulate for electrons came the next year.

Einstein, in fact, took enough interest in Bose's work to follow up with two papers of his own extending Bose's ideas. In particular, Einstein predicted that a gas of noninteracting atoms—material particles—could, at low enough temperature, condense into what later was called a Bose-Einstein condensate.

Back when classical physics ruled, physicists assumed that atoms in a gas, even though in some sense identical—for instance, all helium atoms—were nevertheless identifiably different. You could have helium atom number 1, helium atom number 2, and so on. You could, in principle, keep an eye on atom number 1 and follow its peregrinations, never losing track of it. How such a collection behaved on average when held at a certain temperature was described mathematically by something called the *Maxwell-Boltzmann statistics,* named after two giants of the late nineteenth century, the British physicist James Clerk Maxwell and the Austrian physicist Ludwig Boltzmann. Their description still works fine when quantum effects are not important, such as for a container of helium gas at room temperature.

Bose and Einstein modified the statistics of Maxwell and Boltzmann by taking into account the fact that according to quantum theory, identical particles are *truly* identical, truly indistinguishable. There is no such thing as labeling one among a group of identical atoms and following its motion, even in principle. The wave nature of matter and the governing role of probability make this impossible. When this true identity is factored in, a new statistics emerges, which has come to be called the *Bose-Einstein statistics.* For a dilute gas at ordinary temperature, the Bose-Einstein statistics differ hardly at all from the Maxwell-Boltzmann statistics (an example of the correspondence principle, discussed back at Question 3). But at sufficiently low temperature the quantum identity of the particles can become important. The predicted Bose-Einstein condensate is something completely foreign to classical expectation.

Recall that Bose and Einstein did their work a year before Pauli proposed the exclusion principle for electrons. Pauli's principle introduced a new twist in the treatment of identical particles. Bose and Einstein assumed that their identical particles could share states of motion. Electrons—and later other particles with half-odd-integral spin—could not. So there are two classes of identical particles. Notable leaders in studying the statistics of groups of particles obeying the exclusion principle were the Italian physicist Enrico Fermi and the British physicist Paul Dirac. Both were in their twenties when they developed what we now call the *Fermi-Dirac statistics*.

Now to the naming. Because there are two different kinds of identical particle, they need names. It was Paul Dirac who suggested that particles obeying Bose-Einstein statistics be called *bosons*. (Perhaps he thought that *Einsteinons* was too big a mouthful, or that Einstein was already famous enough). Then he, Dirac, a famously modest man, suggested that particles obeying the Fermi-Dirac statistics be called *fermions*. His coinages have stuck.

Interactions

50. What is a Feynman diagram? Richard Feynman was a brilliant, playful, endlessly curious, much-admired physicist who earned his doctorate from Princeton University in 1942, made major contributions to the Manhattan Project in World War II while still in his twenties, and, after that war, became a world leader in theoretical particle physics. His 1965 Nobel Prize in Physics (shared with the American Julian Schwinger and the Japanese Sin-Itiro Tomonaga) recognized his contributions to what is called *quantum electrodynamics* (or *QED*), the study of the interaction of photons with charged particles, especially electrons and their antiparticles, positrons (see the footnote on page 16). Part of his legacy is the Feynman diagram, a sketch in spacetime of what is happening at the most fundamental level when particles undergo decay or interact with one another.

Space diagrams are quite familiar. They are maps. Spacetime diagrams are less familiar, but not hard to grasp. Figure 24 depicts a map (a space diagram) showing the path of an airplane flight due east from Geneva to Ascona in Switzerland, followed by an automobile trip through the mountains to Berne (where Einstein lived and worked in his famously productive year, 1905). For good measure, the counterrotating beams of protons in the Large Hadron Collider are depicted near Geneva, not to scale. The altitude of the flight isn't shown, nor the depth of the protons' paths, nor the ups and downs of the drive. The lines on the map are what mathematicians call *projections onto a two-dimensional surface*.

Richard Feynman (1918–1988), shown here in 1962. An astonishingly versatile, curious, and fun-loving physicist, Feynman contributed to the understanding of supercomputers as well as superconductivity and fundamental particles. His books include the influential *Feynman Lectures on Physics* and the best-selling *Surely You're Joking, Mr. Feynman.* As a member of the Presidential Commission investigating the Space Shuttle Challenger disaster, he theatrically demonstrated at a press conference the weakness of an O-ring by showing how it became brittle in ice water. Photo courtesy of AIP Emilio Segrè Visual Archives, Segrè Collection.

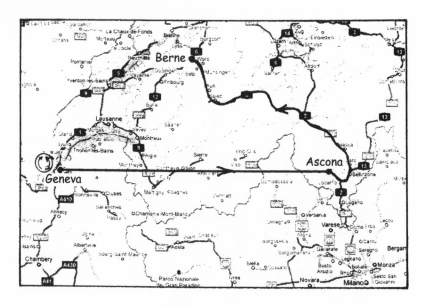

FIGURE 24 A space map.

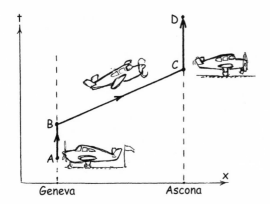

FIGURE 25
A spacetime map.

Now let's switch to a spacetime map, just for the Geneva-to-Ascona flight (Figure 25). In this map, there is only one space direction, east-west, represented by a horizontal axis. There is also only one time direction, represented by a vertical axis. When the plane is sitting stationary in Geneva before takeoff, its "path" in this diagram is a straight vertical line. The plane is not moving in space, but is moving through time. As it flies to Ascona, its spacetime path is a line tilted away from the vertical. Then the plane is moving through both space and time. The greater its speed, the greater the tilt of this line. Finally, after the plane lands in Ascona and parks at the (small) terminal, its spacetime path is again a vertical line. Points in such a spacetime map are called *events*. Several such points are labeled in the diagram. The lines themselves are called *world lines*. Events of interest in such diagrams are usually places where a world line changes in some way, such as when the plane starts moving or stops moving. (In reality, of course, the changes at points B and C are not abrupt.) One other point to notice in the diagram: There are arrowheads on the world-line segments, showing the direction in which events evolve: A to B to C to D. These arrowheads are, in a sense, redundant, for there is no other direction in which the events could evolve. The plane can't fly from C to B, for that would mean going backward in time. In our large-scale world, events unroll inexorably in only one direction in time—the upward direction in the spacetime map. But—and you are unlikely to be surprised by this—things are different in the quantum

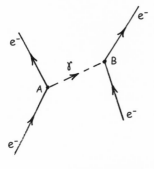

FIGURE 26
Electron-electron
scattering with exchange
of one photon.

world. At least for small enough bits of spacetime, particles can move backward as well as forward in time.

Feynman realized that small spacetime maps for events in the subatomic world could provide a handy way of illustrating and cataloging what is going on when particles interact. In Feynman's hands, these diagrams did more: They provided guides to calculating the probabilities of different events. Here I want to emphasize only their utility as pictorial representations of fundamental events.

Figures 26 and 27 show two Feynman diagrams (among an infinite number of possible ones) for the interaction of one electron with another one. Classically, two electrons interact by a repulsive electric force that "reaches out" from one to the other. Electrons that start toward each other along certain paths and with certain speeds are deflected, retreating from each other along different paths and probably with different speeds. Their interaction has been smooth and continuous. Quantum-mechanically, the situation is completely different. The simplest thing that can happen is shown in Figure 26. Beginning at the bottom of the diagram, you see the world lines of two electrons moving toward one another. At point A, the electron on the left emits a photon (or gamma ray, thus the symbol γ) and changes direction, flying off to the left. At point B, that photon is absorbed by the second electron, which also changes direction, flying off to the right. Viewed "globally," this is not very surprising. Two electrons approached each other, interacted, and rebounded. Viewed "microscopically" (actually, submicroscopically), however; this is quite surprising. The interaction occurred not via a force that reaches out across space, but by the exchange of a photon, emitted at one point and absorbed at another. We say that the deflection of the electrons is the result of an "exchange force." The photon is a "force carrier." The actual interactions occur at the spacetime points A and B.

Figure 27 shows another possibility: that not one but two photons are exchanged. The overall result is the same—electron deflection—but the mechanism is a bit more complicated. As you might surmise from these two figures, there is no end to the ways in which a pair of electrons can exchange photons, and no end to the number of Feynman diagrams that can describe electron deflection. Quantum physics tells us that in fact all possible processes of this kind can and do occur, but with different probabilities. At high energies the simplest processes predominate, so that Figure 26 and 27 pretty much represent what is really going on.

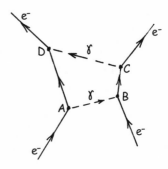

FIGURE 27
Electron-electron scattering with exchange of two photons.

51. What are the essential features of Feynman diagrams? Figures 26 and 27 reveal two important characteristics that show up in every Feynman diagram involving leptons, quarks, or both—which is to say, every interaction that directly correlates with laboratory observation.* First is that at every point of interaction (points A, B, C, and D in the figures), two fermion lines and one boson line are joined. The fermions are fundamental particles (leptons in these figures, but they could also be quarks, as illustrated in later diagrams). The bosons are the force carriers that join them. Those interaction points in the figures are called *vertices*, and, in particular, *three-prong vertices*. The stunning generality that we now believe to be true is that every quark and lepton interaction in the universe is, ultimately, the result of three-prong vertices in which two fermion world lines and one boson world line meet.

The other notable characteristic is that *nothing survives an interaction vertex.* (Look ahead to Figure 30 to see this more plainly illustrated.) At each point of interaction, particles are created and annihilated. Particles that arrive at the vertex are destroyed, to be replaced by new particles

---

*Certain "hidden" interactions, such as between one gluon and another, don't obey the rules set forth here.

that leave the vertex. In Figures 26 and 27, it might appear that an electron arrives at a vertex, emits a photon, and then continues on its way. What the mathematics of quantum physics tells us is that at vertex A, for example, an electron is annihilated and an electron is created. There is no way to say that they are the same electron, for all electrons are identical. Theory (backed up by experiment) supplies us with another stunning generality: All interactions in the universe involve the creation and annihilation of particles.

Figures 26 and 27 contain arrowheads on the particle world lines depicting their forward motion through time. But quantum theory holds another surprise for us. Backward-in-time motion is also possible in the particle world. Consider the Feynman diagram in Figure 28 illustrating the annihilation of an electron and a positron to form two photons, represented by the reaction equation

$$e^- + e^+ \rightarrow 2\gamma.$$

In the diagram an electron arriving from the left and a positron arriving from the right approach each other. At point A the incoming electron is annihilated, and a photon and electron are created. At point B the newly created electron and the positron meet and are annihilated as another photon is created. First, ignore the arrowheads on the lines and imagine a horizontal ruler slid slowly upward from the bottom of the diagram, mimicking the passage of time. First the ruler intercepts the world lines of an electron moving in from the left and a positron moving in from the right. After the interactions, as the ruler approaches the top of the figure, it intercepts the world lines of two photons flying apart—in short, a before and after matching the reaction equation just above.

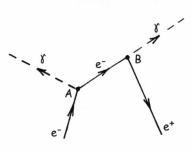

FIGURE 28
Feynman diagram for electron-positron annihilation.

But there is another way to interpret the diagram: An electron emits a photon at point A, from where an electron moves on to point B, from

John Wheeler (1911–2008), shown at age twenty-three when he was a postdoctoral researcher with Niels Bohr in Copenhagen. Later, in 1939, Bohr and Wheeler wrote the definitive paper on the mechanism of nuclear fission, published on the day Hitler's army invaded Poland. Wheeler's special flair was reflected both in his drawings, often produced ambidextrously at a blackboard, and his coinages, which include *Planck length*, *quantum foam*, and *black hole*. I was among a small army of young physicists mentored by Wheeler. Photo courtesy of AIP Emilio Segrè Visual Archives, Wheeler Collection.

where an electron reverses course through time, its line extending downward and to the right from point B. The three fermion lines can be viewed as a single line with two kinks in it. The arrowheads in the figure illustrate this point of view. The idea that a positron moving forward in time and an electron moving backward in time are really the same was first advanced by John Wheeler, Feynman's advisor at Princeton. Here is how Wheeler describes this epiphany in his autobiography:* "Sitting at home in Princeton one evening [in 1940 or 1941] it occurred to me that a positron could be interpreted as an electron moving backward in time. I was excited enough about the idea to phone my graduate student Richard Feynman at the Graduate College, the on-campus residence where he lived. 'Dick,' I said, 'I know why all electrons and all positrons have the same mass and the same charge. They are the same

---

*John Archibald Wheeler, *Geons, Black Holes, and Quantum Foam: A Life in Physics*, with Kenneth Ford (New York: W. W. Norton, 1998), page 117.

particle!'" Feynman liked the idea. Particle world lines moving with equal facility forward and backward in time are a standard feature of the diagrams that bear Feynman's name. Also, the point of view makes for an even simpler definition of an interaction vertex for material particles: a point where one fermion line ends, another fermion line begins, and a boson line begins or ends.

52. How do Feynman diagrams illustrate the strong, weak, and electromagnetic interactions?
Figures 26, 27, and 28 show the electromagnetic interaction at work. The carrier of the electromagnetic force is the photon. What triggers its emission or absorption is electric charge. So anything that carries electric charge, such as an electron or a quark, experiences the electromagnetic interaction. An uncharged object, such as a neutrino, does not. The electromagnetic interaction is intermediate in strength between the strong and weak interactions (actually, much closer to the strong than to the weak*).

Enrico Fermi, when he first developed a theory of the weak interaction in 1934, envisioned a four-prong interaction, as exemplified by the radioactive decay of a neutron, in which one particle transforms to three others:

$$n \rightarrow p + e^- + \bar{v}_e.$$

Although he didn't know about the intermediate process in this decay that came to light later, he did introduce the deep idea that interactions involve annihilation and creation of particles.

In the 1940s, after the muon was discovered to decay "slowly" (in two microseconds) into an electron—and presumably also a neutrino and an antineutrino—physicists assumed that another four-prong vertex was at

---

* In our present "cold" universe, the great disparity in mass between the force carriers of the weak and electromagnetic interactions causes these two interactions to have very different strength (the larger the mass of the force carrier, the weaker the force). In the extremely hot early universe, moments after the Big Bang, the disparity in mass made little difference because most of the energy was kinetic energy, not mass energy. Then the two interactions were of comparable strength.

work, and they spoke of a "universal Fermi interaction." The muon decay is represented by the reaction equation

$$\mu^- \rightarrow e^- + \nu_\mu + \bar{\nu}_e.$$

Figures 29 and 30 show how this process was envisioned initially and how we see it now. The weak interactions are mediated by the very heavy W and Z bosons (whose great mass is part of the reason that the weak interaction is weak). Figure 30 shows a negative muon, initially at rest (vertical line in the figure), transforming first into a muon neutrino and a W^- particle at point A. Then, at point B, the W^- transforms into an electron and an electron antineutrino. So in this Feynman diagram, you see only three-prong vertices, none with four prongs, and you also see illustrated the rule that every vertex involves an incoming fermion, an outgoing fermion, and a boson. (Note that the electron antineutrino that is produced is described as a neutrino moving backward in time.)

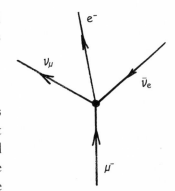

FIGURE 29
Muon decay as originally imagined.

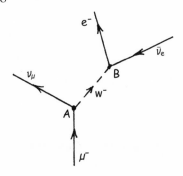

FIGURE 30
Muon decay as now imagined.

The W^- particle in this process "carries" the force, but is nowhere to be seen in the "outside world" (in the above reaction equation, for instance, the W^- makes no explicit appearance). It is created and annihilated in a distance far too small and a time far too short to be measured. It is what is called a *virtual particle*, one that pops into and out of existence. Its effects are real enough, and there is no doubt of the role it plays, but in this process it escapes notice. If you think about energy conservation, you might even wonder how it could take part in the process at all. The total energy available is just the mass energy of the muon, far less than the energy needed to create a W particle. But that is the nature of

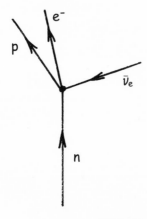

FIGURE 31

Neutron decay as originally imagined.

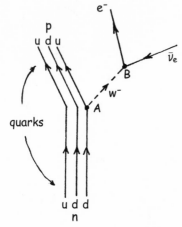

FIGURE 32

The new view of neutron decay.

virtual particles. Heisenberg's uncertainty principle (see Questions 42 and 74) allows for transitory violation of energy conservation in processes that endure a short enough time. One more oddity of quantum physics.

I should add that real W particles, as opposed to virtual ones, have been created and studied in the laboratory. They emerge from particle collisions at the great energies available at the largest accelerators.

Figures 31 and 32 show the evolution of thinking about neutron decay, an even more dramatic change of viewpoint than that for muon decay. Figure 31 presents the original four-prong version of neutron decay, represented by the reaction equation

$$n \rightarrow p + e^- + \bar{\nu}_e.$$

This is the same as the equation displayed at Question 36, except that now the subscript "e" is added to show that the antineutrino is of the electron type.

Figure 32 shows several refinements. First, the quark constituents of the neutron and proton are displayed. They are, respectively, up-down-down (udd) and up-up-down (uud). Second, the decay is shown to occur when one of the down quarks is transformed into an up quark and a W^- particle. Third, the W^- particle is a virtual particle with only a transitory existence, just as in muon decay. Here, too, the available energy is by far insufficient to make a real W^- particle. In this figure, as in Figure 30, points A and B are weak-interaction vertices where a carrier of the weak force, the W^- particle, comes into and goes out of existence.

For the strong interaction, gluons are the force carriers. Note that the electromagnetic interaction has just a single force carrier, the photon; the weak interaction has two force carriers, the W and Z particles (with two possible signs for the W); and the strong interaction has eight force carriers, the gluons. Gluons have unwieldy names like *red-antiblue*, *blue-antigreen*, and so on.* What a gluon does when it is emitted or absorbed is change a quark of one color into a quark of another color. Figure 33 shows a Feynman diagram for a nucleon that is just "sitting still."

Because of gluon exchanges, the quarks are constantly changing color, but there remain three of them, with their up-down definitions not changed. Note that every vertex, in keeping with the general rules, is a point where one fermion is destroyed, one is created, and a boson is either created or destroyed. A motionless nucleon is a lively place.

Theory says that at sufficiently high energy the picture of Figure 33 can be transformed into a maelstrom of quarks and gluons, called a *quark-gluon plasma* (see Question 92).

In this answer I have skipped over the gravitational interaction. It, like the other three interactions, is, in principle, mediated by a boson. The quantum carrier of the gravitational force is called a *graviton*. Like the photon, the graviton is massless, but it has 2 units of spin instead of the photon's 1 unit. Because of the extreme weakness of gravity, there is to date no evidence for the graviton. It remains a theoretical entity. Gravitons act together in such vast

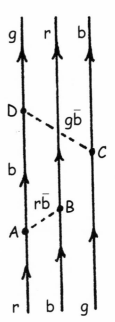

"Colored quarks"

FIGURE 33
Quarks exchanging gluons.

---

* As mentioned on page 123, if you count up how many color-anticolor combinations you can get from three colors, you are likely to arrive at nine. For technical mathematical reasons, only eight are independent.

multitudinous numbers that there is scant hope of seeing evidence for just one or a few.

53. Which particles are stable? Which are unstable? What does it mean to say that a particle decays?

It's a pretty good approximation to say that *every* particle is unstable (or, what is the same thing, radioactive). Every particle will decay into particles lighter than itself (downhill in energy) unless something prevents it from doing so. So far as we know, the only things that can prevent it are a pair of conservation laws (laws of constancy): the law of quark number conservation and the law of charge conservation. Without those two laws, all particles in the universe would dissolve into a sea of photons and neutrinos (and perhaps gravitons). There would not be many scientists to study the matter.

From the law of quark number conservation it follows that baryon number is conserved. The stabilization of the proton, in turn, can be attributed to baryon conservation, for the proton is the least massive baryon. It can't decay because there is no less massive baryon into which it might decay. This seems to be an absolute conservation law—although it is always dangerous in science to call anything absolute. There are in fact some theories that suggest that the proton *might* decay. If it does, its mean life is unimaginably long. It is known to have a mean life greater than 2×10^{29} years (or greater than 10^{32} years for particular assumed modes of decay). The lesser of those limits is about 10 billion billion times the lifetime of the universe (that is, the time since the Big Bang), which is a mere 1.4×10^{10} years. This might suggest that if the proton is unstable, there would be no way to measure its half-life. This is not necessarily true, because of the way probability works in quantum physics. Suppose, for example, that the half-life of the proton is 10^{30} years. Half of the protons in a piece of matter will live longer than that (if the universe lives that long!). But some will, by chance, decay sooner, a very few *much* sooner. How much matter would you need to assure an average of one decay per year? About three tons should do it.*

---

*This calculated requirement assumes that neutrons bound within nuclei decay at the same rate as protons.

What the law of charge conservation does is stabilize the electron, for the electron is the least massive charged particle. If it decayed, it would have to be into neutral (uncharged) particles. The law of charge conservation, like the law of baryon conservation, seems to be absolute, or very nearly so. A current lower limit on the mean life of the electron is 5×10^{26} years—not as long as the limit on the proton life but still *very* long. If this were the lifetime of the electron, you would expect to see, on average, one electron decay per year in a kilogram or so of material.

These lower limits on the proton and electron mean life are not based on infrequent observations of decay. Rather, they are based on *not* seeing any decay events.

There is one other prohibition at work in particle decays—the prohibition against decay into a single particle. Thus a muon can't decay into an electron and nothing else, nor a lambda into a neutron and nothing else, even though such decays would be consistent with the conservation of energy, charge, and baryon number. What gets in the way of one-particle decays is momentum conservation. Momentum, a property of motion determined by both mass and speed, must be the same before and after a decay event. So if a lambda at rest were to decay into a neutron only, the neutron would have to remain at rest (zero momentum before = zero momentum after). But in that case energy would not be conserved, since the neutron's mass is less than the lambda's mass. If energy *is* conserved in this hypothetical one-particle decay, the neutron that is produced would fly away with some kinetic energy, but then there would be momentum after the event that was not there before the event. So, no one-particle decays. (A physics student can prove that this rule holds true even if the initial particle is moving.) In practice, then, every unstable particle decays into two or more other particles. The Feynman diagrams earlier in this section illustrate that rule.

54. What is scattering? Decay is what can happen, and usually does happen, to a single particle. Even though that particle does not appear to be interacting with anything else, in fact the decay is brought about by some interaction, often the weak or electromagnetic interaction.

Decay is to be distinguished from scattering. When two particles collide, they can bounce off each other or they can be destroyed while other particles are being created. Such processes are more obviously the result of interaction and are called *scattering,* whether the final particles are the same as or different from the initial particles. Strictly speaking, the particles that result from scattering are *always* different, even if they are the same (please excuse the Alice-in-Wonderland quality of that statement). In Figure 26, for instance, two electrons come together and two electrons fly apart. But what happens at vertex A (and also at B), according to the mathematics of quantum theory, is that an electron is annihilated and an electron is created. It is meaningless to say that the electron that disappears is the same as the electron that appears, but it is *also* meaningless to say that they are surely different. All we can say is that they are electrons, and all electrons are identical. How, then, can we think about it without having our heads spin? The thinking that best matches the mathematics is that the newly created electron is distinct from the one that is destroyed.

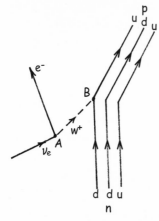

FIGURE 34

A neutrino (perhaps from the Sun) interacts with a neutron to produce an electron and a proton, a process observed at Sudbury Neutrino Observatory.

According to the definitions just given, Figure 28 also depicts a scattering process. It shows an electron and a positron that collide and are destroyed as two photons are created. Another example of scattering, this one involving quarks, is shown in Figure 34. An electron neutrino collides with a neutron, resulting in an electron and a proton:

$$\nu_e + n \rightarrow e^- + p.$$

This is a process observed at SNO, the Sudbury Neutrino Observatory in Canada (see Question 41). The incoming neutrino comes from the Sun. The neutron that is struck is held—that is, stabilized—within a deuteron (single neutrons do not sit around in the laboratory;

they decay). So what actually happens is that a neutrino strikes a deuteron, creating an electron and two protons. But one of the protons is a "spectator," so I leave it out of the diagram. As Figure 34 shows, a W^+ particle acts as the intermediate "force carrier." Where it is created at point A, a neutrino is annihilated and an electron created. Where the W^+ particle is destroyed at point B, a down quark is transformed into (or replaced by) an up quark.

55. What is the same before and after a scattering or a decay? A

quantity that is the same after a particle process as before it is called a *conserved* quantity, and the law stating its constancy is called a *conservation law*. A few of these you have encountered already. There are quite a few conservation laws, some seemingly absolute—that is, valid for any and all processes—and some that we call *partial*. A partial conservation law holds for some interactions but not all. Conservation laws, the very bedrock of particle processes and indeed of large-scale processes as well, are the subject matter of Section X. Here I just draw attention to some of the conserved quantities that show up in the Feynman diagrams in this section.

Energy conservation is one of the absolutes. It explains why decay processes are always "downhill" in mass and why a lot of kinetic energy is required in collisions to make new mass. I have also mentioned momentum conservation, which explains why a particle can't decay into a single other particle.

A third "mechanical" conservation law is that of angular momentum, or spin. Among its consequences is that if a system has half-odd-integral spin before an interaction, it also has half-odd-integral spin after the interaction. Correspondingly, integral spin is preserved from before to after. These rules are illustrated in the decay of the muon, Figure 30, and the annihilation of an electron and positron, Figure 28. Because spin is a vector quantity, one has to do a bit more than just count. For instance, in the decay of a muon, a particle of spin $\frac{1}{2}$ gives rise to three particles, each of spin $\frac{1}{2}$. You have to think of two of the three final particles having their spins oppositely directed, and so canceling, leaving a final total spin of $\frac{1}{2}$. Similarly, in the electron-positron annihilation, one possibility is that the two initial particles have oppositely directed spins, for a total spin of zero. Then the two photons that are created, each of spin 1, must also have oppositely directed spins, for a total of zero.

More evident in the diagrams are the conservation of "intrinsic" properties such as charge, lepton number, and baryon number. If you examine Figures 26 through 34, you will notice that charge, lepton number, and baryon number (a quark's baryon number is $\frac{1}{3}$) are conserved not only from before to after, but separately at each interaction vertex. I said earlier that nothing is preserved at an interaction vertex. I should have said that no *particle* is preserved. At the vertex, particles come into and go out of existence, but certain properties of the particles are not lost. Electric charge or lepton flavor are a bit like a baton in a relay race. One runner "disappears," another runner "appears," but the baton continues on, emerging from the handoff, the point of "interaction."

Figure 33 shows one other conservation law at work, that of color. At point A, for instance, a red quark disappears, giving way to a blue quark, but at the same time a red-antiblue gluon is created. So one unit of "redness" is preserved from before to after at that interaction point, and zero unit of "blueness" is also preserved, since blue and antiblue together have zero "blueness," just as an electron and positron together—a lepton and antilepton—have zero lepton number. Similarly, at point C, there is 1 unit of "greenness" before and after and zero unit of "blueness" before and after. Terrible nomenclature, to be sure, but it's the best that physicists could come up with.*

56. What changes during a scattering or decay? A short answer is that anything that is not conserved may change—mass, for example. It is *total* energy, including mass energy, that doesn't change, but mass may decrease, as it does in decay events, or increase, as it often does in scattering events. Another thing that can change is the number of bosons.

---

* It is said that some American physicists wanted quark and gluon colors to be red, white, and blue, but they yielded to the international consensus for red, green, and blue.

When an electron and positron annihilate to form two photons, for instance, the number of bosons before the process is zero, and after the process it is two (see Figure 28). And the number of particles of a specific type, such as negative muons or down quarks, may change. In Figure 30, for instance, you see that a negative muon does not survive the interaction, but nevertheless the number of particles in the muon *flavor* is preserved, for the muon is replaced by a muon neutrino. And in Figure 32, showing the decay of a neutron, the number of up quarks changes and the number of down quarks changes, but the total number of quarks does not change.

Despite the fact that quite a few things can change in interactions, the conservation laws provide powerful restrictions limiting what can happen. This will be evident in the next section.

section X

Constancy during Change

57. What are the "big four" absolute conservation laws? In the world around you, it appears that nothing is static. Clouds shift, leaves flutter, you move from place to place. As Euripides put it in the fifth century B.C., "All is change." It's not hard to agree with that sentiment. And indeed, the focus of most scientists looking at nature in the twenty-five centuries since Euripides has been on change. That was Newton's focus in the seventeenth century when he studied how things move in response to forces. It was Maxwell's focus in the nineteenth century when he studied the emission of radio waves by oscillating electric charge.

But amid all the change, *some* things do remain constant. Those are the conserved quantities. We now understand that things known to remain constant in the large-scale world (the "classical" world) are also constant in the quantum world—in fact, right down to the level of individual acts of particle interaction. The *reason* that charge is conserved in the large-scale world is that it is conserved in every instance of particle annihilation and creation. The *reason* that energy is conserved in the large-scale world is that it is conserved in the quantum world. If you think about it, it's not obvious that this tight link between the large- and small-scale worlds *has* to be true. Suppose, for example, that a kindly grandmother always keeps exactly one hundred cookies in a cookie jar. Every time the grandchildren visit, the jar contains the same number of cookies. As fast as they are removed, they are replaced. But that particular cookie conservation law doesn't arise because individual cookies are conserved. Cookie conservation is a law in the large, but not in the small.

However, for the "big four" of classically conserved quantities—energy, momentum, angular momentum, and electric charge—we do believe that the conservation laws hold down to the tiniest domains of space and time, and that, at all scales, these laws are absolute (insofar as we dare say that anything in the universe is absolute).

A simple example that shows all four conservation laws at work (and actually more) is the decay of a positively charged pion into a positively charged muon (which is actually an antimuon) and a neutrino of the muon flavor (see Figure 35):

$$\pi^+ \to \mu^+ + \nu_\mu.$$

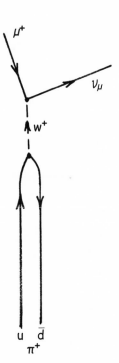

Energy conservation is reflected in the "downhill" character of the decay. The masses of the product particles add up to less than the mass of the pion, so mass energy decreases. The muon and neutrino then fly apart with enough kinetic energy to balance the energy books. *Momentum conservation* is reflected in *how* the product particles fly apart. Suppose that the pion is initially at rest, patiently awaiting its inevitable end. Then its momentum is zero. Accordingly, the total momentum of the muon and neutrino must also be zero. These two particles fly apart back-to-back with equal momentum. By the rules of vector addition, the two oppositely directed momenta add to zero. *Angular momentum conservation* also involves vector addition (since angular momentum, like ordinary momentum, is a vector quantity). The spin of the pion is zero. The muon and neutrino each have ½ unit of spin, which, if oppositely directed, can also add to zero. *Charge conservation* is more evident in the reaction equation. The total charge is 1 unit before and after the decay.

FIGURE 35
In this example of pion decay, numerous quantities are conserved.

You can look back to any of the reaction equations displayed earlier in the book to see that they are consistent with the big four conservation laws.

58. What additional absolute conservation laws operate in the quantum world? There are three other conservation laws and a symmetry principle that have shown up only in the subatomic domain and are also believed to be absolute. They are listed, along with the big four, in Table 1. Two of the new conservation laws are actually illustrated by the pion decay discussed above. Lepton flavor is preserved in that decay because it is zero both before and after the decay. It is zero before because then only the pion is present—no leptons. It is zero afterward because a lepton of the muon flavor and an antilepton of the same flavor are created. (Note that it is the neutrino that is the lepton in this case and the positive muon that is the antilepton.) When a negative pion decays, on the other hand, its products are a negative muon and an antineutrino of the muon flavor. It is just a matter of convention that negative leptons are called *particles* and positive leptons *antiparticles*. That choice arose from the preponderance of electrons over positrons in our world. Had the reverse choice been made, the physics would have been the same.)

The pion decay also illustrates the conservation of quark number. The positive pion is constructed of an up quark and an antidown quark. By the usual rules for counting particle numbers, this adds up to zero quark number. After the decay, there are no quarks, so the number remains zero.

The third new absolute conservation law listed in the table is of *color,* a property of quarks and gluons only. Color can be visualized as

Table 1 **Quantities Believed to Be Absolutely Conserved**

Energy	Lepton flavor* (implies lepton number)
Momentum	Quark number (implies baryon number)
Angular momentum	Color
Charge	TCP (an invariance principle)

*Lepton flavor is "almost" absolutely conserved—seemingly in all instances *except* neutrino oscillation.

being something like electric charge—and indeed it is sometimes called *color charge*. It is a "certain something" carried by a particle and passed on whenever that particle interacts. As I have mentioned before, the three colors are conventionally called *red, green,* and *blue.* Quarks carry color, antiquarks carry anticolor, and gluons carry a mixture of color and anticolor. There are several ways in which an entity can be colorless. It can contain no quarks or gluons at all—an electron, for instance, or a neutrino. Or it can be an equal mixture of a color and its anticolor—red and antired, for instance. The positive pion I have been discussing is a combination (a superposition) of red-antired, green-antigreen, and blue-antiblue quarks, making it colorless. And, finally, an equal mixture of the three colors, with no anticolors, is colorless. It may help to remember this if you recall that equally mixing the colors of the rainbow produces white light—although, in truth, there is no connection between the two kinds of color mixing. Figure 33 on page 147 displays the colorless combination of three quarks in a nucleon. If you think of time running upward in the figure, you can place a ruler horizontally at the bottom of the figure and move it up slowly (all the while keeping it horizontal) to simulate the passage of time. You will note that at every instant, the left, middle, and right quarks have different colors, so the colorless combination red-green-blue is preserved moment to moment throughout the quark-gluon dance.

The last in the list of new absolutes in the table is a symmetry principle called the *TCP theorem.* It is listed along with conservation laws because it, too, provides an ironclad rule on what may happen in nature.

59. What is the TCP theorem? Each of the letters T, C, and P stands for a kind of reversal, real or hypothetical. Here, in brief, are the meanings of the three reversals:

> T, *time reversal:* Run the experiment backward—that is, interchange before and after.
>
> C, *charge conjugation:* Replace all particles in an experiment with their antiparticles and all antiparticles with their particles.
>
> P, *parity, or mirror reversal:* Run the experiment as the mirror image of the original experiment.

What the "TCP theorem" states is that if all three of these reversals are applied to any process that actually occurs, the triply reversed process is also one that can occur, and indeed in every particular.

Until 1956, physicists believed (without good evidence, as it turned out, so it was a matter of faith—and I was among the faithful) that T invariance, C invariance, and P invariance were separately valid laws. Instead they have turned out to be partial conservation laws. I will discuss their "partiality" in answer to the next question. Here let me expand a bit on the nature of these three reversals and how they combine to make an absolute law.

T: To imagine one time-reversed process, look at the reaction equation for positive pion decay on page 155 and read it from right to left. A positive muon and a muon neutrino collide to produce a positive pion. Or go back to Figure 28 on page 142 and read the Feynman diagram from top to bottom: Two photons collide to produce an electron and a positron (these are both highly unlikely, but not impossible, processes). In the large-scale world, a time-reversed process would be what you see when you run a video or movie film backward. In the ordinary world of our experience, time-reversed processes are usually impossible (and often look ridiculous) because of the astronomically large number of particles involved. Think, for example, of droplets of water rising from blades of grass to join in a stream that flows back into the nozzle of a hose. Or the shattered remains of cars that had crashed reassembling themselves into shiny vehicles as they back away from the scene of the crash. On the other hand, some processes in the large-scale world are simple enough that it would be hard to tell the difference between the original process and its time-reversed image. Think of a baseball flying from third base to first or a satellite circling Earth.

C: To stick with the positive pion decay that I have been discussing, you can easily see the result of charge conjugation, or particle-antiparticle interchange. The C-reversed process is the decay of a negative pion (which is the antiparticle of a positive pion) into a muon and a muon antineutrino:

$$\pi^- \rightarrow \mu^- + \bar{\nu}_\mu.$$

Indeed, a negative pion decays in this fashion. *But*—and here was the big surprise that shattered physicists' faith in the validity of C invariance alone—the decays do not mimic each other in every particular, as C invariance implies. The reason is that the neutrino is "left-handed" and the antineutrino "right-handed." Figure 36 shows what is meant by left-handed. If the thumb of the left hand points in the direction of the neutrino's motion, the curved fingers of the left hand indicate the direction of the neutrino's spin. Applying C-reversal to the positive pion decay leads to what is in fact an impossible process, the decay of a negative pion into a muon and a left-handed antineutrino.

FIGURE 36
The neutrino is a left-handed particle.

P: Parity reversal is the same as looking at the mirror image of a process. If you look at any page in this book in a mirror, you see backward writing that might be hard to read. Yet you know that what you are seeing is not impossible to achieve. One could set backward type (as is often done on the front of ambulances) so that the mirror writing looks normal. Perhaps as a child you did "mirror writing" so that your parents or siblings could read what you wrote only by looking at it in a mirror. For some particles, on the other hand, the mirror view is not the view of something that is possible. The mirror view of a left-handed neutrino is a right-handed neutrino, and they don't exist (Figure 37).

Finally, let's apply all three reversals to the positive pion decay, which I now rewrite with subscript L for the left-handed spin of the neutrino:

$$\pi^+ \to \mu^+ + \nu_{\mu L}.$$

T changes before to after. C changes particles to antiparticles (and vice versa). P changes left-handed spin to right-handed spin. The triply reversed process is

$$\mu^- + \bar{\nu}_{\mu R} \to \pi^-.$$

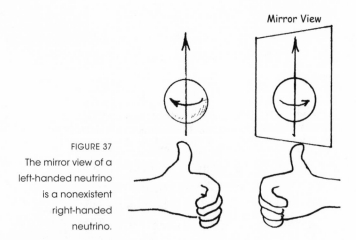

FIGURE 37
The mirror view of a
left-handed neutrino
is a nonexistent
right-handed
neutrino.

Although there would be no hope in practice of making this process occur, we have every reason to believe that it is a possible process. TCP invariance seems to be an absolute law.

60. What conservation laws are only "partial"? It might seem at first that a conservation law can be no more partial than a pregnancy. A quantity either is or isn't constant. But something can be conserved in one interaction and not in another. Such a quantity is said to be partially conserved. Table 2 shows two such partially conserved quantities, isospin (which I will explain in a moment) and quark flavor, plus the "almost" absolutely conserved quantity lepton flavor. Quark flavor is preserved in strong and electromagnetic interactions, but not in weak ones. Isospin is preserved only in strong interactions; it is violated by both electromagnetic and weak interactions. You can perceive a rule here. The strong interaction is hemmed in by the most conservation laws, the electromagnetic interaction by fewer, and the weak interaction by still fewer. This raises an interesting question, to which no one knows the answer: Does the gravitational interaction, the weakest of all, violate still more conservation laws? If the answer to this question is eventually found to be yes, it will mean that some of the laws we now call *absolute* are not really so.

The concept of isospin dates from the discovery of the neutron in 1932. The neutron's mass was found to be very close to that of the proton,

Table 2 **Partially Conserved Quantities**

Isospin (except in weak and electromagnetic interactions)
Quark flavor (except in weak interactions)
Lepton flavor (except in neutrino oscillation)

and their spins are the same. Early evidence suggested that the neutron and proton experienced near-identical strong interactions within nuclei. This gave rise to the idea that the neutron and proton are two different "states" of the same underlying entity, the nucleon. Werner Heisenberg (the same Heisenberg who, a few years earlier, had been a chief architect of quantum mechanics and had offered the uncertainty principle) devised a theory in which "flipping" a nucleon from its proton state to its neutron state is mathematically identical to flipping an electron spin from up to down. For the electron, there are just two possible directions of its spin. For the nucleon, there are just two possible states, neutron and proton. Thus the name *isospin*, even though the "flipping" back and forth between neutron and proton, has nothing to do with actual spin.

The nucleon, in Heisenberg's theory, is a "doublet." Later, other "multiplets" were found, including a triplet of three pions, a doublet of two xi particles, and a singlet lambda particle. When a multiplet is made up of two or more particles, they have very similar masses but different charges. The law of isospin conservation says that all the members of any multiplet experience identical strong interactions. Because the members of the multiplet have different charges, it is clear that they do not experience identical electromagnetic interactions. They may also differ in their weak interactions. Thus isospin is a partially conserved quantity.

Since isospin is a relevant quantum property only for strongly interacting particles, its conservation can be incorporated into quark conservation laws. Nevertheless, it is worth a separate discussion because of its long history and the fact that it deals with observable particles, not just the unobservable quarks.

The second partial conservation law listed in Table 2 is that of quark flavor. Recall that each of the six quarks is said to have its own flavor. The tongue-twisting flavors (thank heaven there are not thirty-one of them) are called *downness, upness, strangeness, charm,** *bottomness,* and *topness.* An example that shows this conservation law at work is the creation of a sigma particle and a kaon when a pion collides with a proton:

$$\pi^+ + p \rightarrow \Sigma^+ + K^+$$

In terms of the quark constituents, this is rendered as

$$u\bar{d} + uud \rightarrow uus + u\bar{s}.$$

This is a strong-interaction process that can happen with high probability in an accelerator (after the pion is created in a prior collision). Upness is conserved, with three up quarks on the left and three on the right. The down and antidown quarks on the left have a total downness of zero, which is preserved on the right. And the zero strangeness on the left is matched by a strange and antistrange quark preserving zero strangeness on the right.

Now consider the decays of the two particles created in this process:

$$\Sigma^+ \rightarrow n + \pi^+$$

and

$$K^+ \rightarrow \mu^+ + \nu_\mu.$$

Expressed in terms of quark constituents, these decays are

$$uus \rightarrow ddu + u\bar{d}$$

and

$$u\bar{s} \rightarrow none.$$

You can see that in the sigma decay, although upness is conserved, downness increases by one and strangeness decreases by one. It's as if a strange quark turned into a down quark. In the kaon decay, upness decreases by one and strangeness increases by one, as if an up quark had turned into a strange quark. These are processes that do indeed occur, but because they

---

*Logic would dictate *charmness,* but logic does not always prevail in particle-physics nomenclature.

don't conserve quark flavors, they occur by the weak, not the strong interaction. Consequently, they are "slow," with half lives of about 10^{-10} second and 10^{-8} second, respectively.

For the record, Table 2 also lists lepton flavor as only partially conserved. It is an "almost" absolute law (and so appears in Table 1) but is violated by neutrino oscillation.

61. What symmetry principles are only "partial"? Table 3 shows that the three "reversal" symmetries are only partial, being violated by the weak interactions, although the three of them taken together (TCP) constitute what is believed to be an absolute symmetry. Here I will discuss how, taken one at a time, they are violated by the weak interactions.

The first suggestion that parity (P) might not be a valid symmetry principle for the weak interactions came in 1956 from a pair of young Chinese-American theorists, Tsung-Dao (T. D.) Lee, then twenty-nine and at Columbia University, and Chen Ning (Frank) Yang, then thirty-three and at the Institute for Advanced Study in Princeton. Lee's Columbia colleague Chien Shiung Wu immediately set out to test the parity principle. Madame Wu, as she was known to all her students and colleagues, enlisted the aid of researchers at the National Bureau of Standards in Washington who were expert in achieving very low temperature. In her experiment she cooled a sample of the radioactive isotope cobalt-60 to within a small fraction of 1 degree of absolute zero in order to line up the nuclei in a magnetic field. (A temperature of even 1 degree would be so "warm" that all of the nuclei would jiggle out of alignment.) Figure 38 shows the arrangement schematically and the results.

In the figure, the fingers of a right hand indicate the direction of the spin of a cobalt-60 nucleus. The upward-pointing thumb indicates that

Table 3 **Partial Symmetry Principles**

T	time reversal
C	charge conjugation (particle-antiparticle interchange)
P	parity, or mirror invariance
PC	combined

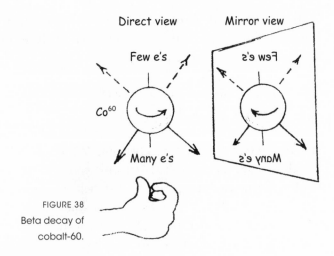

FIGURE 38
Beta decay of
cobalt-60.

the "north pole" of the nucleus is on top. What Madame Wu found was that when many such nuclei are lined up, the electrons from their beta decay are shot out preferentially downward (in the "south pole" direction), only few emerging upward. Now what does this process look like in a mirror? As shown in the figure, the mirror view reveals a nucleus with its north pole on the bottom and with electrons shooting out preferentially in that direction. The mirror view is a view of the impossible. If parity symmetry were a valid principle, electrons would have to fly away equally in the upward and downward directions, for only then would the mirror view and the actual view be equivalent. It is so simple and yet so profound. (Getting to within three thousandths of a degree of absolute zero was the big challenge in this experiment—now routine but then a tour de force.)

Shortly after Madame Wu's work, another Columbia University team showed that *both* C-invariance and P-invariance are violated by the weak interaction. Leon Lederman (later to receive a Nobel Prize for other work), along with Richard Garwin and Marcel Weinrich, used the Columbia cyclotron to create positive pions and then studied the decay of the pions into muons (and unseen neutrinos). The central panel in Figure 39 shows a positive pion decaying into a positive muon (that is to say, an antimuon) and a muon neutrino. These researchers found, by studying the subsequent decay of the muon, that it is "single-handed." At

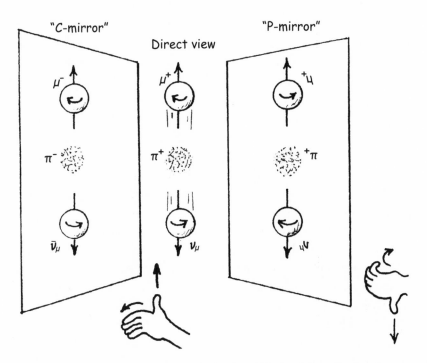

FIGURE 39 Decay of positive pion, with "C-mirror" and "P-mirror" views. The thumbs show spin directions of the neutrino in the "C-mirror" and "P-mirror." Other straight arrows designate direction of travel.

the time, they could not tell if the muon was left- or right-handed, but they could infer, correctly, that the unseen neutrino must also be single-handed, with the same handedness as the muon. This is because of momentum and angular momentum conservation. Momentum conservation requires that the two decay particles fly apart back-to-back. Angular momentum conservation requires that the total spin of the decay particles be oppositely directed so that their sum (their vector sum) is zero, matching the pion's zero spin. In the figure I show the positive muon to be left-handed (as it was later learned to be). If your left thumb points in the direction in which the muon is moving (upward), the curved fingers of your left hand show the muon's spin direction. Since the neutrino is moving in the opposite direction *and* spinning in the opposite direction, it, too, is left-handed.

Now come the arguments about P and C. On the right in the figure is an ordinary mirror, here called a *P-mirror*. That mirror shows a positive pion decaying into a right-handed positive muon and a right-handed neutrino. This is an impossible process, since neutrinos are left-handed. P is not a valid symmetry. On the left is a "C-mirror," which magically interchanges particles and antiparticles and leaves other properties unchanged. It shows a negative pion decaying into a left-handed negative muon and a left-handed antineutrino—again, an impossible process, since antineutrinos are right-handed. So C is not a valid symmetry. (Keep in mind that these statements about P and C apply only to the weak interaction, which governs this decay process.)

For some time after these experiments were concluded, physicists considered it likely that the combined CP invariance was valid, even if C and P separately were not. If that were true (it isn't), then T invariance would be an absolute principle. But it was not to be. In 1964, two Princeton University physicists working at the Brookhaven accelerator on Long Island, Val Fitch and James Cronin, discovered that a long-lived version

Val Fitch (b. 1923), shown here on the October morning in 1980 when he had just been notified of his Nobel Prize. Fitch started life on a cattle ranch in Nebraska. As a young soldier in World War II with only part of an undergraduate education behind him, he was assigned to work at Los Alamos, where he rubbed shoulders with many of the great physicists of the time and learned techniques of experimental physics. As a Columbia University graduate student, he pioneered the study of "muonic atoms," measuring the energies of photons emitted by negative muons as they cascaded through energy states around heavy nuclei. When I got to know Fitch in the 1990s, he had the same informality and charm as suggested in this photo. Photo by Bill Saunders; courtesy AIP Emilio Segrè Visual Archives.

of the neutral kaon, although it usually decays into three pions, decays occasionally—about one time in five hundred—into two pions. Theory showed that this could happen only if combined CP invariance, and, with it, T invariance, are not absolutely valid principles. This discovery, which earned Fitch and Cronin the Nobel Prize in 1980, was even more unsettling to physicists than the earlier upending of P and C invariance. The reason is that it implies a basic difference between matter and antimatter. Val Fitch likes to say that the failure of PC (and T) invariance is the reason that we are here. Physicists now reckon that because of the lack of perfect symmetry between matter and antimatter, the early universe, shortly after the Big Bang, contained a not quite equal number of quarks and antiquarks. For every 1 billion quarks, according to calculations, there were 999,999,999 antiquarks. When the dust cleared, one quark out of each billion survived, to make protons, neutrons, galaxies, stars, planets, and us.

62. What are laws of compulsion and of prohibition? Quantum physics brought with it two big changes in the way we perceive and interpret the basic laws of nature. One is that the smooth continuity of the large-scale world has given way to explosive events of annihilation and creation in the subatomic world. The other is that certainty has been replaced by probability. How can two such different world views be compatible? Because when you pile those annihilation-creation events one on top of another by the trillion, multiple explosions simulate continuity and probabilities move toward certainty.

A typical classical law of physics can be called a *law of compulsion*. It dictates what *must* happen once initial conditions are specified. For example, if you loft a spacecraft to some point well above Earth's atmosphere and turn off its engine, the gravity of Earth, Moon, Sun, and planets will guide it inexorably along some path through space. If you know its location and its velocity at one moment, you can calculate with assurance where it will be forever after, under the compulsion of Newton's laws of motion. (There will remain a bit of uncertainty because you don't know its initial location and velocity with absolute precision).

Another example from the classical world: If you set the electrons in an antenna of given length and given orientation to vibrating with a certain amplitude and a certain frequency, you can calculate the exact pattern of electromagnetic radiation that will be broadcast, under the compulsion of Maxwell's laws of electromagnetism.

By contrast, quantum laws are more likely to tell you what *can't* happen than what *must* happen. In that sense, they are laws of prohibition. If one proton strikes another at a certain energy, for example, there may be numerous possible outcomes. Here are some possibilities for an available energy of 1 GeV or so:

$$p + p \rightarrow p + p$$
$$\rightarrow p + p + \pi^0$$
$$\rightarrow p + p + \pi^+ + \pi^-$$
$$\rightarrow p + n + \pi^+$$
$$\rightarrow \Sigma^+ + n + K^+$$

If there are many such proton-proton collisions, all identical, these and various other possibilities will be realized, each with its own probability. On the other hand, the list of outcomes that will *not* occur is limitless. Any outcome that violates any absolute conservation law will not occur. (If an outcome violates only a partial conservation law, it may occur, but with much reduced probability.) The conservation laws are the police that prohibit. Most evident here is the conservation of charge. Any outcome in which the total charge is not +2 is prohibited. Other prohibited outcomes are any in which the energy increases, the number of quarks changes, a single lepton is created, or the spins of the final products can't combine to zero or a small integral value.

This range of possible outcomes raises an interesting question: Do all possible outcomes that are consistent with the conservation laws actually occur? Physicists think that the answer to this question is yes, although it isn't obvious that the answer has to be yes. For instance, if there are four supermarkets within a convenient range from your home, all selling products that you like at prices you consider reasonable, it doesn't follow that you shop at all of them, each with a certain frequency. For whatever reasons, there might be some of them at which you never shop. There are a vast number of "prohibited" supermarkets, those outside your commuting range or incompatible with your preferences, and only four "allowed" supermarkets. As a classical creature, you decide which among those four will get your business. If you were a quantum creature, you would have no choice: You would shop at all four—although not equally often. That all of the allowed possibilities do seem to occur leads some physicists to bring the idea of compulsion back into the picture by saying, "Everything that *can* happen *must* happen."

I alluded before to "initial conditions," such as the location and velocity of a spacecraft when turned loose. A head-on collision of one proton with another at a certain energy is also an initial condition. The difference between classical laws and quantum laws is this: Classically, only one outcome is realized for a given set of initial conditions. Quantum mechanically, there may be many possible outcomes for a given set of initial conditions. *Conservation laws* allow certain outcomes and prohibit others. *Quantum laws* dictate the relative probability of the allowed outcomes.

63. How are the concepts of symmetry, invariance, and conservation related? Physicists are sometimes a bit casual in using the terms *symmetry*, *invariance*, and *conservation* interchangeably. They mean different things, although they do have something in common: They all deal with the absence of change—with something that remains constant. Moreover, the three ideas are related to each other.

Even when physicists are trying to be careful, they do not always mean the same things when they refer to symmetry, invariance, and conservation. Here are some workable definitions: *Symmetry* occurs when some change, real or hypothetical, leaves the *physical situation* unchanged. *Invariance* exists when some change—also real or hypothetical—leaves one or more *laws of nature* unchanged.* *Conservation* occurs when some specific *physical quantity* remains the same during a process of change. Let me clarify through example.

---

* *Invariance* can also mean that some calculated quantity remains unchanged, but I shall not expand here on that usage.

FIGURE 40

Types of symmetry.

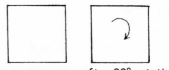

Same appearance after 90° rotation

A square (see Figure 40) has a kind of rotational symmetry. After a rotation through 90 degrees (or 180 or 270), its appearance is the same as before the rotation. A straight railroad track with evenly spaced ties has what is called *translational symmetry*. Any shift along its direction by the distance between ties or any whole multiple thereof leaves its appearance unchanged. An Ingres portrait of a woman's face may have left-right symmetry. This means that if the left and right side of the portrait are interchanged, the appearance is unchanged, or nearly so. The mask in Figure 40 shows this symmetry more exactly. Left-right symmetry you will recognize to be the same as mirror symmetry, or the parity transformation.

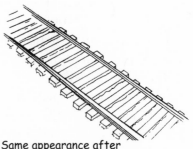

Same appearance after
shifting an integral number of ties

Same appearance after
left-right interchange

When I discussed the "reversal symmetries" P, C, and T in answer to Question 61, I actually went beyond symmetry and moved into invariance, because there the question was not just "Do things appear the same after the reversal," but "Are the reversed physical processes possible." So the question was whether the laws of nature are the same ("invariant") after the inversions of parity, charge conjugation, and time reversal are applied.

The symmetries I have just been discussing—of the square, the railroad track, and the mask, along with P, C, and T—are "discontinuous symmetries." It takes a finite change to restore conditions to what they were before. We also recognize certain "continuous symmetries." A chief one of these is the homogeneity of space. One way to put this is that in remote empty space, you don't know where you are and have no way of finding out. One part of space is exactly like every other part. This is true whether you shift by a nanometer or a mile. Space itself has no distinguishing characteristics. It is everywhere the same. This symmetry (a displacement in any direction by any amount leaves the physical situation unchanged) leads directly to an invariance principle: that the laws of nature are the same everywhere. We have every reason to believe that this is true; there is abundant evidence of it in what we are able to learn about the motions of distant galaxies, nuclear reactions in far-away quasars, and the passage of photons across the universe.

Another continuous symmetry is the "isotropy of space." This means that there is not only no preferred position in space, there is also no preferred direction. The related invariance principle is that the laws of nature are independent of orientation.

In 1915 there were not many women mathematicians in the world, and not many of either gender with the insight of Emmy Noether (pronounced, roughly, NUR-ter, with the first R suppressed). In that year, at age thirty-three, working at the University of Göttingen in Germany under the "protection," so to speak, of the famous mathematician David Hilbert,* she devised an important theorem linking symmetry and conservation. Still known as *Noether's theorem*, it states that for every continuous symmetry in nature there is a conservation law. (The theorem applies to classical as well as quantum physics.)

Without using her mathematical approach, let me try to explain how the homogeneity of space (its sameness everywhere) is related to the conservation of momentum, one of the implications of Noether's theorem. Imagine a spacecraft in remote outer space—truly remote, free of all gravitational influences—moving in some direction with some velocity.

---

* Despite Hilbert's best efforts on her behalf, Noether did not gain a regular faculty position at Göttingen until 1919. A course that she taught in 1916–1917 was listed in the university catalog as "Mathematical Physics Seminar: Professor Hilbert, with the assistance of Dr E Noether, Mondays from 4–6, no tuition."

According to Newton's first law of motion, we know that if we look at the spacecraft later, it will be somewhere else but moving in the same direction with the same velocity. Newton's first law states that in the absence of outside influences, an object moves in a straight line with constant speed. Its momentum doesn't change. Digging deeper, we can ask *why* this is true, and the answer is that space is homogeneous. If the spacecraft spontaneously changed its momentum, it would reflect some difference in space between where it was and where it is. Or, if it were initially motionless and then started to move, it would be observed later at a different point with a momentum that it didn't have before.

It is a mind-bending idea that the bland sameness of space from one place to another can account for such a vital pillar of physics as the law of momentum conservation.

Or consider a compass needle in outer space, where there is no magnetic field to deflect it. If at one moment it points in some direction, we do not expect it to turn spontaneously to point in some other direction. Were it to do so, it would imply that not all directions in space are equivalent—that space is *not* isotropic. It turns out that when Noether's theorem is applied to the directional symmetry of space, the implication is that angular momentum is conserved. This fundamental law that controls all particle interactions springs from the fact that all directions in space are equivalent.

Just for completeness, I mention that the conservation of energy rests on time symmetry—the physical equivalence of any time to any other time.

Notice that these continuous symmetries are also related to invariance principles. The laws of nature don't depend on where you are or in what direction you are moving or what time it is.

section XI

Waves and Particles

64. What do waves and particles have in common? How do they differ? At first thought, it might seem that waves and particles have next to nothing in common. Waves are spread out over space and have nebulous boundaries. You can't say that a wave is "at that spot." Particles, by contrast, are little nuggets with well-defined boundaries—or perhaps, if they are leptons or quarks, with no physical extension at all. Particles have mass. We don't usually assign mass to a wave. Waves are characterized by wavelength, frequency, and amplitude, concepts that have no obvious counterpart for particles. Waves can *diffract* (bend around corners) and *interfere* (reinforce or cancel one another), behaviors not expected of particles. Indeed, the demonstrated interference of light in the early nineteenth century convinced scientists that light is made of waves, not particles.

Yet, if you think about what waves can *do,* you begin to see things they have in common with particles. Waves, like particles, can possess energy and momentum and can transmit these quantities from one place to another. Like particles, they move at a certain speed. Moreover, waves can be at least partially localized—that is, to have boundaries, even if the boundaries are not quite sharp. The vibrating air in an oboe is more confined than the air in a bassoon, which in turn is more confined than the air in a large organ pipe. Or think of yourself and a friend holding the two ends of a stretched-out rope (see Figure 41). If you snap your end,

FIGURE 41 A wave pulse, like a particle, carries energy and momentum from one place to another.

a localized wave pulse runs down the rope. It transfers an impulse to your friend's hand, just as if you had thrown a ball to her.

The marriage of wave and particle came with Einstein's corpuscle of light (the photon). It was, for a long time, a union blessed by very few scientists. From 1905 until the mid-1920s, the idea that light could be both wave and particle was too much for most scientists to swallow. Compton's 1923 experiments on the scattering of X-rays by electrons in atoms convinced some. Bose's derivation of Planck's radiation law convinced others. Proof positive came in 1927 when Clinton Davisson and Lester Germer, working at Bell Labs in the United States, and, independently, George Thomson (son of J. J., the electron's discoverer), working at Aberdeen University in Scotland, found that electrons striking the surface of a solid crystalline material exhibited diffraction and interference effects (see Figure 42). From the known spacing of atoms in the crystal and from the measured angle at which electrons preferentially bounced from the crystalline surface, they could even measure the wavelength of the electrons.

In these early experiments, the wavelength of the electrons was comparable to the spacing between atoms in a solid. Later, experimenters learned how to slow down neutrons so much that their wavelength greatly exceeded the spacing of atoms in a solid. (I will explain this inverse connection between speed and wavelength in answer to the next question.) As a result, a neutron drifting lazily through a material "reaches out" via its wavelength to interact simultaneously with many atoms, behavior hardly expected of a particle smaller than a single atomic nucleus.

65. What is the de Broglie equation? What is its significance? By chance, the young French Prince Louis-Victor de Broglie (pronounced,

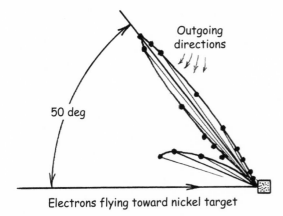

Outgoing
directions

50 deg

Electrons flying toward nickel target

FIGURE 42

Experimental results of Davisson and Germer showing that electrons of 54 eV, after striking a nickel crystal, emerge mostly in a certain direction because of diffraction and interference of the electron waves. Image adapted from *Nobel Lectures, Physics* (Amsterdam: Elsevier, 1965).

roughly, *"Broy"*) received his bachelor's degree in physics in 1913, the same year as Niels Bohr's groundbreaking work on the quantum theory of the hydrogen atom. I don't know if quantum physics was yet on de Broglie's mind at that time, but later, in his Nobel Prize address of 1929, he did speak of his attraction to "the strange concept of the quantum, [which] continued to encroach on the whole domain of physics."

Following service in World War I, de Broglie took up graduate study in physics and, in 1924, as part of his doctoral dissertation at the University of Paris, he offered the deceptively simple but powerful equation that now bears his name. The de Broglie equation is written

$$\lambda = h/p$$

On the left, the symbol λ (lambda) stands for wavelength—evidently a wave property. The p in the denominator on the right stands for momentum—clearly a particle property. Linking the two is Planck's constant h, which appears in *every* equation of quantum physics. This equation was known to be true for photons—if you believed in photons. De Broglie asserted it to be true for electrons and all particles. When Davisson, Germer, and Thomson measured the wavelengths of electrons a few years later, they found that indeed their measurements conformed to de Broglie's equation. The de Broglie equation has stood the test of time and remains a pillar of quantum physics. It is as simple in appearance and, in its way, as powerful as Einstein's $E = mc^2$.*

De Broglie said later that two things led him to his equation, which expresses what we now call the *wave-particle duality*. One was the evidence supplied by Compton's then-recent 1923 work that X-rays exhibit particle as well as wave properties. The other was de Broglie's observation that in the classical world, waves are often quantized but particles, apparently, are not. He was thinking of the fact that the strings on a violin and the air column in a flute vibrate with only certain selected frequencies, not arbitrary frequencies. He wondered whether the quantization then known to exist in atoms might be explained as the result of a vibrating wave—whether, in effect, the atom is a musical instrument.

The structures of Einstein's and de Broglie's equations differ in a simple yet significant way. In $E = mc^2$, the E and the m are both "upstairs." E is directly proportional to m. More mass means more energy. In $\lambda = h/p$ by contrast, λ is "upstairs" and p is "downstairs" (in the denominator). Wavelength and momentum are *inversely* proportional. This means that increasing a particle's momentum decreases its wavelength. So protons in more powerful accelerators have shorter wavelength than protons in less powerful accelerators, making them better able to probe phenomena at subnuclear distances. Neutrons slowed to very small speed—and small momentum—acquire relatively long wavelengths that enable them to reach out to interact with many atoms at a time. There are implications on the human scale, too. When you walk down the street, you have momentum. If de Broglie is right, you must have a wavelength as well. Where is it? Why don't you experience it? It is there, but because your momentum is so enormous on an atomic scale, your wavelength is too small by many orders of magnitude to detect. A 150-pound person strolling at two miles per hour has a wavelength of 4×10^{-34} inch—not encouraging for measurement. "But," you might say, "I can move more slowly, in order to decrease

---

* You may wonder why Einstein's equation does not contain Planck's constant h. It is because $E = mc^2$ is basically a *classical* equation, which happens to be valid in the quantum world as well.

my momentum and increase my wavelength." Good thought, but it's hopeless. If you creep at 1 inch per century, your wavelength will be 5×10^{-23} inch, far less than the diameter of a single proton. We humans, like it or not, are denizens of the classical world.

But when you shrink the mass enough to enter the particle world, wavelength becomes very significant indeed. Because of its wave nature, an electron within an atom spreads out to encompass the whole atom. Similarly, neutrons and protons within the nucleus spread themselves over the nuclear volume. Only when a particle is accelerated to great energy does its wavelength shrink to less than the size of a nucleus or even the size of a single neutron or proton. Then the high-energy particle, with its shrunken wavelength, becomes a good probe of the smallest distances.

66. How are waves related to quantum lumps? When de Broglie offered his idea that all particles have wave properties, the full theory of quantum mechanics was yet to appear (it was developed in the following two years, stimulated in part by de Broglie's work). In 1924, scientists still imagined electrons in atoms to be tracing out "Bohr orbits." De Broglie's ingenious idea was to suppose that the *reason* only certain orbits were allowed was the self-reinforcement, or constructive interference, of the wave that accompanied the particle in its orbit. As indicated in Figure 43, a wave could—and usually would—interfere destructively with itself after completing one trip around the orbit. Then, as de Broglie imagined it, the wave would simply wipe itself out, averaging to zero after multiple trips around. Such an orbit would simply not exist. But for certain selected orbits, the circumference would be an exactly integral number of

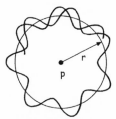

Wave destructively
interferes with itself

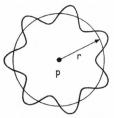

Wave contructively
reinforces itself

FIGURE 43
De Broglie's idea of
a wave interfering
with itself.

wavelengths, and the wave, after completing one or any number of trips around, would constructively reinforce itself. That, suggested de Broglie, would explain why only certain orbits were permitted within the atom. The orbiting electron would be like the stretched string on a guitar, capable of vibrating with only certain selected wavelengths. Quantization in the atom then becomes no more mysterious than the "quantization" of vibrational frequencies in a musical instrument.

When de Broglie applied this reasoning to circular orbits in the hydrogen atom, he found exact agreement with the experimentally known energies in that atom. The lowest-energy state, or "ground state," was the one in which one wavelength stretched around the circumference, the first excited state had two wavelengths stretched around the circumference, the next state three wavelengths, and so on. The soon-developed quantum mechanics upended this picture of the atom, but not totally. De Broglie's idea of self-reinforcing waves within atoms survived the revolution. Yet there were big changes. The biggest, perhaps, was that the electron is not to be looked at as a particle accompanied by a wave. Rather, it *is* a wave—one spread over the whole three-dimensional space within an atom, not just stretched out along an orbit. Consequently, the oscillation of the electron wave is not just circumferentially around the nucleus, it is also radially in and out toward and away from the nucleus, and can be a combination of both the roundabout and in-and-out vibrations.

Figure 44, for example, shows two possible distributions of electron intensity (really of electron probability) for the fourth state (the third excited state) in a hydrogen atom. The shaded regions show where the electron is likely to be found if something—such as a high-energy X-ray—probes the atom in search of a particle amid the wave. In the upper diagram, the high-intensity regions are concentric circles (with an intensity peak also right at the nucleus). This state of motion is one with zero angular momentum, meaning that the electron, instead of circling around the nucleus, can be visualized as running back and forth radially, toward and away from the nucleus. It is analogous, is a way, to the waves on the surface of still water that spread out radially when a pebble is dropped into the water. The direction of propagation of such a wave is away from the center, not around the center.

By contrast, the lower diagram shows the wave pattern for a state of motion with the most angular momentum that is possible at this energy (actually, three units). This pattern corresponds to circular motion around the nucleus and thus matches de Broglie's original idea. It is an example with four full wavelengths around the circumference. The reason you see eight dark regions in the diagram rather than four is that a peak of intensity occurs at every minimum as well as every maximum of the wave, so there are two peaks per wavelength. Notice in the diagram that the dark regions, although concentrated at a certain radius, do have some in-and-out spread. You can never completely corral a wave.

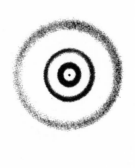

FIGURE 44

Two possible wave distributions for the third excited state in the hydrogen atom, one with the electron running radially in and out, the other with the electron circling the nucleus.

Just a reminder of how these diagrams relate to energy quantization. There can be two or three or four wavelengths in and out or roundabout, but never two-and-a-third wavelengths or four-and-a-half wavelengths. Because of the restriction to a whole number of wavelengths, there is a restriction to only certain energy values, just as de Broglie originally hypothesized.

Back at Question 3 I discussed the correspondence principle, the idea introduced by Niels Bohr that when a quantum property changes by a small fraction from one state to the next, quantum behavior closely mimics classical behavior. This is illustrated in Figure 45, which shows the electron intensity pattern in the twentieth state of the hydrogen atom with the greatest possible angular momentum for that energy. You see, very clearly, a wave following a circular path far from the atomic nucleus. There are forty intensity peaks, corresponding to the twenty full wavelengths around the circumference, and there is a lot of empty space (actually, *almost* empty space) between the orbit and the nucleus. The intensity is bunched up radially and the pattern begins to resemble a classical circular orbit—although, to be sure, still showing a clear

wave pattern. A little closer to the nucleus there is an allowed state of motion with nineteen wavelengths around its orbit, and a little farther out there is one with twenty-one wavelengths around its orbit. These permitted orbits, although quite distinct, each with its own energy, are, percentagewise, not far apart, again illustrating the correspondence principle, since classically an orbit could have any radius at all.

67. How do waves relate to the size of atoms? Imagine that you have a proton in one hand and an electron in the other. You release them at some distance apart with only empty space between them. What happens? They are drawn together by an electric force (since unlike charges attract). Classically, the electron would either dive straight into the proton or spiral around it in ever smaller loops as it radiates away energy. That classical expectation is not unlike what happens when you release a marble within a bowl. It will either roll straight down, eventually settling at the center as friction dissipates its energy, or it will spiral downward, still ending up at the center. It will have found the point of lowest energy. The electron would "like" to do the same, but its wave nature gets in the way. As the electron orbits closer and closer to the nucleus, its wave "pulls in" to an ever shorter wavelength. Enter de Broglie. Shorter wavelength means larger momentum. As the space over which the electron ranges gets smaller, its wavelength gets shorter, and it moves ever faster as its momentum grows. You see the momentous significance of the *inverse* relationship between wavelength and momentum.

And as momentum grows, kinetic energy grows. As the electron gets confined to less and less space, its energy of motion gets larger and larger. It is almost as if there is a repulsive force countering the attractive electric force. In energy terms, there are two competing effects. Electric attraction lowers the energy—the *potential* energy—as the electron gets closer to the nucleus, while, at the same time, the electron's wave nature causes an increase in energy—its *kinetic* energy—as it gets closer to the nucleus. The two effects find a point of balance where the total energy is a minimum—a point where moving in toward the nucleus would cause the total energy to increase as the kinetic energy effect dominates and moving away from the nucleus would also cause the total energy to increase as the potential energy effect dominates. That point of balance for a single electron and proton is at a distance of about 10^{-10} m (a tenth of a nanometer), very large by usual particle standards—in fact, about one hundred thousand times larger than the size of the proton itself. Thanks to the wave nature of matter, atoms are "huge"—huge relative to the size of any composite particle such as a proton, neutron, or pion.

The small mass of the electron also plays a role in fixing the relatively large size of the atom. Momentum is the product of mass and speed, so it is not only small speed but also small mass that contributes to large wavelength. Because the electron is far less massive than any other particle (except the photon and neutrinos), it has, typically, less momentum and greater wavelength than other particles and so occupies more space. (The neutron provides an exception to this rule. It can be slowed so much that despite its large mass it can have a large wavelength and, as I mentioned before, spread itself over a space as large as that occupied by several atoms.)

There are two other questions related to the wave nature of the electron and the size of atoms. One is why heavy atoms, despite having more powerful electric attractive forces, do not collapse to a size much smaller than the size of a hydrogen atom. The other question is why an atom in an excited state can be quite a bit larger than in its ground state. Here I amplify answers to these questions given at Question 31.

If you considered the fate of a *single* electron near, say, a uranium nucleus, it would indeed find its point of least energy, its point of balance, much closer to the nucleus than in a hydrogen atom. But as you add more and more electrons, they feel a less and less powerful force, on average. The ninety-second electron in the uranium nucleus is pulled inward by ninety-two protons and pushed outward by ninety-one other electrons. The net pull on it is about the same as in a hydrogen atom, so that last electron settles into a state of motion not very different in size from the size of a hydrogen atom.

As to why excited atoms are larger than ground-state atoms: The wave for an electron in its lowest state has just one cycle of oscillation. It rises and falls just once over a dimension that defines the size of that state of motion. In an excited state, the electron wave undergoes two or more cycles of oscillation. So the excited-state wave needs more "elbow room" to complete its multiple cycles of oscillation. That is a bit of an oversimplification but it gives the general idea. If the excited-state wave, with its multiple oscillations, were squeezed down to a size as small as the ground-state wave, it would have a shorter wavelength, a greater momentum, and a greater kinetic energy. It could then lower its total energy and find a point of balance at a larger size.

68. What is diffraction? What is interference? There is something basically "fuzzy" about waves. They don't follow narrowly defined paths like the paths of baseballs or spacecraft. They occupy a region of space, have no well-defined boundaries, spread as they propagate, and can overlap one another.

When a wave passes through an opening or by an edge, it bends (Figure 46). That is called *diffraction*. It can be seen in water waves that pass a ship at anchor, or it can be experienced indirectly by the fact that your wireless phone usually works even if there is a building between you and the cellular antenna. The diffraction effect is more pronounced for larger wavelength, which explains why longer-wavelength AM radio signals bend around obstacles more readily than shorter-wavelength FM signals do. Driving in the canyons of a big city, you are likely to find AM stations to be a bit more reliable than FM stations.

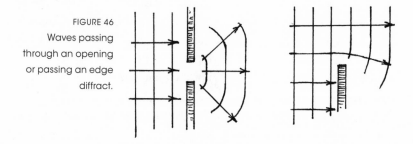

FIGURE 46
Waves passing
through an opening
or passing an edge
diffract.

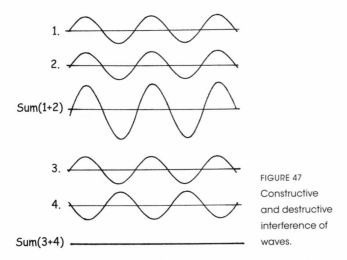

1.

2.

Sum(1+2)

3.

4.

Sum(3+4)

FIGURE 47
Constructive
and destructive
interference of
waves.

When waves overlap (see Figure 47), one possibility is that they line up crest to crest and trough to trough. Then each wave enhances the other and the net result is a wave of greater amplitude. Or they could line up crest to trough and trough to crest, in which case they can wipe each other out (if they are of equal intensity) or produce a wave of less amplitude than either one alone. Or something in between. Physicists refer to all of these possibilities as *interference*. Interference can range from "constructive" to "destructive."

Diffraction and interference were known for some kinds of waves, notably water waves, long before light was recognized for sure to have a wave character. By observing diffraction and interference for light in 1801, Thomas Young established once and for all that light consists of waves, not particles. Or so it was thought. Waves can diffract and interfere, but particles cannot, went the reasoning. Now, of course, we know that particles can diffract and interfere, too. But to give waves their due, particles behave this way only because they have a wave character. De Broglie had it right.

69. What is the two-slit experiment? Why is it important? Young made use of two holes in an opaque screen when he demonstrated the diffraction and interference of light. Now we use slits, which let through

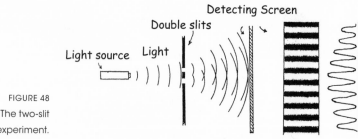

FIGURE 48
The two-slit
experiment.

more light (or more photons). The essence of the two-slit experiment is shown in Figure 48. Light of a single wavelength (monochromatic light) illuminates a screen containing two narrow, closely-spaced slits. Because of diffraction, the light doesn't simply project the image of the two slits on a detector screen. Instead, light passing through each slit is bent this way and that, spreading onto a broad swath on the detecting screen. Any given point on the detecting screen will be receiving light from both slits, which opens the possibility of interference.

The point on the detecting screen that is exactly opposite the midpoint between the two slits is equally distant from the two slits. This means that at that point light from the two slits will be "in phase," crest to crest and trough to trough. Constructive interference is the result. That spot will display a bright band of light. If one moves a little way up the screen from that point, a point will be reached where the distance from the lower slit is exactly half a wavelength greater than the distance from the upper slit. At that point the two waves will be arriving crest to trough and trough to crest. Destructive interference is the result. There a dark band will appear. Moving up a bit farther will bring one to a spot where the distance from the lower slit is one whole wavelength greater than the distance from the upper slit. Again, constructive interference. Again, a bright band of light.

And so on. The overall result is a series of parallel bright and dark bands, sure evidence of diffraction and interference. (From the measured spacing of the bands and the known spacing of the slits and their distance from the detecting screen, one can calculate the wavelength of the light. Light of greater wavelength—red, for instance—will produce more widely separated bands than light of shorter wavelength—blue, for instance.)

This experiment has been carried out with a light source so weak that no more than one photon is anywhere between the source and the detecting screen at any one time. It has also been carried out with electrons, and even with atoms. The result is always the same. A resulting pattern of light and dark bands shows that diffraction and interference have occurred and that the individual particles exhibit a wavelength.

Some physicists say the two-slit experiment is the quintessential quantum experiment because it is so simple in concept and at the same time it so starkly illustrates the wave-particle duality. Consider, for example, the one-photon-at-a-time version of the experiment. One photon passes through the apparatus and lands somewhere on the detecting screen. But where? How does it "decide" where to land? It is guided only by probability. There are some places where it has a good chance to land, other places where it is less likely to land, and some places where it has no chance to land. Figure 49 shows a simulation in which the landing points of one, ten, one hundred, one thousand, and then ten thousand photons are shown (with each panel including the results of the panel above).* For up to ten photons no pattern at all can be discerned. The points look random. For one hundred photons one begins to see a pattern of darker and lighter bands. At one thousand photons the pattern becomes more evident, and at ten thousand photons the pattern conforms to the classical expectation of light and dark bands of constructive and destructive interference.

How is one to interpret the results of the one-photon-at-a-time experiment? How does each photon know where to land—and where not to land? Does it know where other photons have already landed? If a photon passes through one slit, how does it know that the other slit is there (for if only one slit were open, the pattern would be quite different)? Einstein called quantum physics spooky. Richard Feynman described the two-

---

*The results are different every time the simulation is done. In another trial, the upper panels (one and ten photons) would look very different, the middle panels (one hundred and one thousand photons) somewhat different, and the bottom panel (ten thousand photons) hardly different at all.

FIGURE 49

Simulation of the points where photons (or other particles) are detected after being fired one at a time at a double slit. The five panels show, respectively, the results for one, ten, one hundred, one thousand, and ten thousand particles. Each panel includes the results of the previous panel. The narrow stripes in the bottom panel are an artifact associated with pixel size. Images courtesy of Ian Ford, online at www. ianford.com/dslit/.

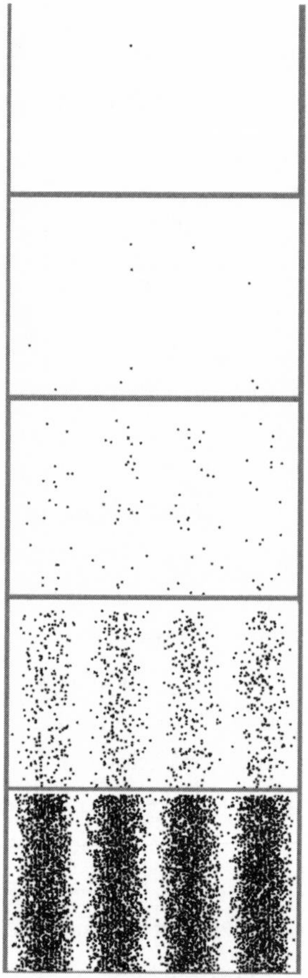

slit experiment as the essence of quantum physics. There appears to be only one way to interpret the experiment: Each photon separately acts as a wave—a wave that passes through both slits—guided solely by the probability associated with that wave, entirely independent of what other photons that went before it may have done.

What the two-slit experiment teaches us—and what myriad other experiments confirm—is that a particle acts as a particle when it is created and annihilated (emitted and absorbed) and acts as a wave in between. To get our heads around it, we just have to give up the idea that a photon is a particle at any moment other than the moments of its birth and death.

70. What is tunneling? In answering earlier questions, I referred to a phenomenon called *tunneling* whereby a particle can penetrate a barrier that, according to classical physics, is impenetrable. Barrier penetration is important in alpha decay, in nuclear fission, in nuclear fusion, and in a modern supermicroscope, the scanning tunneling microscope. The wave nature of matter provides a way to understand tunneling.

Figure 50 shows a potential-energy wall—a prison wall, so to speak. A particle on its left side has insufficient kinetic energy to surmount the

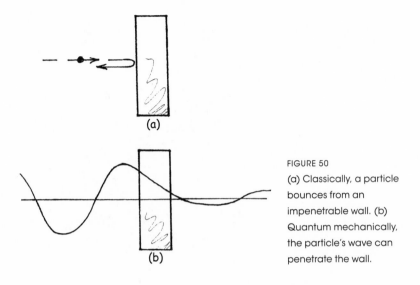

(a)

(b)

FIGURE 50

(a) Classically, a particle bounces from an impenetrable wall. (b) Quantum mechanically, the particle's wave can penetrate the wall.

wall. Classically, as shown in part (a) of the figure, the particle bounces from the wall, and can do so any number of times. It will never get through. It is imprisoned forever. But quantum-mechanically, the situation is a little different because, as shown in part (b) of the figure, the particle's wave can penetrate into the wall. Depending on the height of the wall and the mass of the particle, the wave might die off quite rapidly within the wall, penetrating very little distance (there is no hope that a real prisoner in a real jail could escape by leaning casually against the wall in the exercise yard). But if the wall is not too high (in energy terms) and not too thick, a particle's wave could penetrate all the way through the wall, with a tiny tail making its appearance outside the wall. This bit of wave on the far side of the wall represents a real probability that the particle can appear there, that it can literally pop through the wall.

As so often happens in physics, a phenomenon discovered in a pure physics setting—in this case, alpha decay—finds practical application. One of the uses of tunneling is in the scanning tunneling microscope, or STM, perfected in 1982 by Gerd Binnig and Heinrich Rohrer,* research-

---

* An achievement recognized rather promptly with a 1986 Nobel Prize.

STM image of a "corral" of iron atoms on a copper surface. Image originally created by IBM Corporation.

ers at an IBM lab in Zurich, Switzerland. In their device, a tiny metal tip is brought close to a solid surface (*very* close, within a nanometer, or 10^{-9} m). An electric voltage is applied between the tip and the surface. Classically no current would flow between them because the air is such a good insulator. But that layer of air acts as an energy wall that can be breached by electron waves. There is a small "leakage" of wave from tip to solid surface, resulting in a small "tunneling current." The thinner the wall—that is, the less the distance between tip and surface—the greater the tunneling current. In practice, through a feedback circuit, the tip is moved slightly up and down as it traverses a horizontal path above the surface, the amount of its up-and-down motion being adjusted to maintain a constant tunneling current, which means a constant distance from the surface. Thus the tip traces out the tiny hills and valleys of the surface. As shown in the photo, it can measure distances appreciably smaller than the diameter of a single atom.

section XII

Waves and Probability

71. What is a wave function? What is Schrödinger's equation? I have referred earlier to the role of probability in quantum physics—for instance, that an electron has a certain chance to be found at various places within an atom, or that a photon passing through a double slit has a certain chance to land at various spots. Something called the *wave function* controls this probability. A wave function gives an *amplitude*, a measure of how strongly a wave deviates from a zero value, either positively or negatively. This is analogous to how much water surface deviates, upward or downward, away from its normal level as a wave passes by. But there is one big difference between the quantum wave and the water wave. The quantum wave is not itself a measurable quantity. The quantity that can be measured is probability, and that is the *square* of the quantum wave amplitude. Moreover, the quantum wave amplitude can be what is called a *complex number,* meaning that it has real and imaginary parts. The probability is then given by what is called the *absolute square* of the amplitude, which is indeed a real number and a measurable quantity. In the discussion that follows, I will not pursue this technical detail. Suffice it to say that the squared amplitude gives the probability.

Probability implies uncertainty, a lack of sure knowledge. Indeed uncertainty, or "fuzziness," is a hallmark of quantum physics. Yet there is definiteness involved here, too. The wave function of an electron in a particular state of motion in a hydrogen atom has a perfectly well-defined value at every point. Its square, the probability, is also perfectly definite. And this definiteness extends from one atom to another. If a thousand hydrogen atoms are all in the same state of motion, they all have identical wave functions and identical distributions of probability. What is *not* definite is advance knowledge of where the electron might be found as a particle if an experiment is conducted to pinpoint its location. Identical experiments carried out on the thousand atoms might reveal the electron's location at a thousand different points. It's a bit like throwing a pair of dice. You know exactly (if it's a fair and balanced pair) the probability of throwing any given sum ($\frac{1}{6}$ for a sum of seven, $\frac{1}{36}$ for "snake eyes," and so on), but you never know in advance the result of any given throw.

In late 1925, little more than a year after de Broglie postulated the wave nature of matter, Erwin Schrödinger, a multitalented Austrian physicist, came up with an equation that still bears his name. It is an equation that permits the calculation of the wave function—either precisely, as for a hydrogen atom, or in principle, for any system.* Schrödinger, thirty-eight at the time, was "elderly" compared with most of the other architects of the new quantum mechanics.†

What Schrödinger's equation does is generalize the de Broglie idea of a link between momentum and wavelength to situations where a particle's

----

* Here is what the Schrödinger equation looks like for a particle moving in one dimension:

$$d^2\psi/dx^2 + (2m/\hbar^2)[E - V]\psi = 0.$$

There is a lot of power locked up in that simple-appearing equation. In it, ψ is the wave function, x is the position coordinate along the direction of motion, m is the particle's mass, $\hbar$ is Planck's constant divided by 2π, V is the particle's potential energy, and E is the total energy. It has always tickled physicists that Schrödinger devised this equation while on a winter holiday with a woman friend in Arosa, Switzerland.

† Here are the ages of some other notable quantum physicists in 1925: Paul Dirac and Samuel Goudsmit, twenty-three; Werner Heisenberg and Enrico Fermi, twenty-four; Wolfgang Pauli and George Uhlenbeck, twenty-five; Louis-Victor de Broglie, thirty-three. In that year, Niels Bohr reached forty and Albert Einstein reached forty-six.

momentum and therefore its wavelength vary continuously from one place to another. It might seem at first that a wave cannot have one wavelength at one place and a different wavelength at a slightly different place. Wavelength, by its very nature, seems to be something that is spread out. The wavelength of a water wave or of a vibrating string or column of air is the distance from one crest to the next. But it is actually possible to define a wavelength over a tiny distance by the "curviness" of the wave. As indicated in Figure 51, a wave with a short wavelength is quite curvy and one with a long wavelength is less curvy. Keeping in mind the inverse relationship between wavelength and momentum, this means that a particle has a very curvy wave function where it has large momentum and a less curvy wave function where it has small momentum.

The original idea of de Broglie that a wave could stretch out around the circumference of a circular orbit proved to have some merit because an electron in a circular orbit has constant speed and constant momentum—just as a satellite in a circular orbit around Earth does. So the electron, with its constant momentum, can have a fixed wavelength. But what of an electron that moves directly toward or away from a nucleus rather than around it? As the electron is drawn closer to the nucleus, it speeds up, gaining momentum. So its wavelength—as it moves, say, from a great distance away to within a short distance of the nucleus— gets continually shorter and shorter. This is where the Schrödinger equation steps in. It provided a solution for that type of motion—motion with zero angular momentum—as illustrated in Figure 52 for the particular case of the ground state. The graph shows a greater curviness of the wave function close to the nucleus, and a lesser curviness farther away.

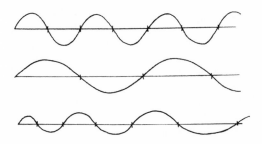

FIGURE 51
Waves of short wavelength are "curvier" than waves of long wavelength. "Curviness" can define the wavelength of any wave at any point.

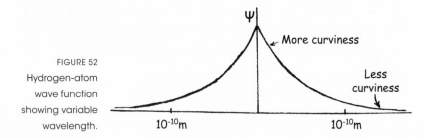

FIGURE 52
Hydrogen-atom
wave function
showing variable
wavelength.

10^{-10}m 10^{-10}m

Two other points about the wave function depicted in Figure 52: First, it rises from a small value to a large value and then sinks back to a small value, so it shows one-half cycle of a full oscillation from one side of the atom to the other. This means that it displays an approximate average wavelength, which is about 0.2 nm (2×10^{-10} m). This, in turn, reveals an approximate average momentum of the electron in this state. Second, the wave nature of the electron means that it can never follow a simple track, either circular or in-and-out. Its wave is always spread out. For the particular state of motion depicted by the graph in Figure 52, you have to think of a fuzzy ball for which every straight-line radial direction is equivalent to every other one. The electron's wave function has a small value around the outer surface of the ball and grows larger as one moves toward the center of the ball. There is no difference between one direction and another.

72. How do waves determine probabilities? I have already stated that the probability of finding a particle somewhere is the square of its wave function at that point. Consider a very simple example of motion: a particle bouncing back and forth with constant speed between two impenetrable walls. The classical view of that motion is shown in both the upper and lower parts of the left panel in Figure 53. Even classically, probability could enter the picture. You might not know exactly where the particle is at some moment until you look. That is *probability of ignorance* (a term I introduced in the answer to Question 27). The particle is surely somewhere, but you don't know where. You just know that it is equally likely to be anywhere. If you take a flash photo of the particle at a hundred random times, you might find it in a hundred different places, scattered more or less uniformly, left to right, over the domain of its motion.

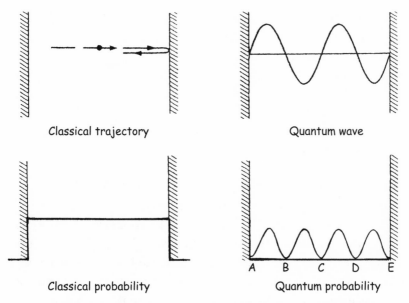

Classical trajectory

Quantum wave

Classical probability

Quantum probability

A   B   C   D   E

FIGURE 53   Classical and quantum descriptions of a particle in a box.

The quantum description of this motion differs in two very important ways from the classical description. First, according to the quantum description, only certain energies are allowed. The energy is "lumpy." A permitted energy is one in which a half-wavelength fits between the walls, or a whole wavelength, or one-and-a-half wavelengths, or any integer number of half wavelengths. Only then will the wave, in multiple trips back and forth, reinforce itself through constructive interference. For any other wavelengths, the wave will, in multiple trips back and forth, destructively interfere with itself, wiping itself out—as originally visualized by de Broglie for circular orbits. The particular wave function shown in the upper right panel in Figure 53 is for the fourth state (third excited state), the one with four half-wavelengths between the walls.*

---

*The Germans have a word for such a wave function: *Eigenfunktion*. This is sometimes translated as "proper function," but more often rendered in English simply as "eigenfunction." In the 1920s, German was a principal language of physics.

The other notable difference between the classical and quantum descriptions is the way in which probability enters the picture. Classically, probability—probability of ignorance—*may* play a role, but it doesn't have to. It depends on what level of detailed knowledge you have of the motion. If you have no sure knowledge of the particle's location, its classical probability is the same everywhere, a constant. This is rendered graphically by the straight horizontal line in the lower left part of the figure. Quantum mechanically, probability *must* play a role. It is fundamental to the quantum description. The lower right drawing in Figure 53 shows the distribution of probability for the illustrated state. Since the probability is the square of the wave function, it can never be negative (fortunately). It takes on a range of values from zero to some maximum. Perhaps the most astonishing feature of that graph is that at five points—at the two walls and at three points in between—the probability is exactly zero. This means that if the particle is in that particular state of motion, it will *never* be found at one of those points. If it is never to be found at points B, C, or D, you might then ask, how could it get from A to E? How could it cover the distance from one wall to the other without having some probability of being found at every point in between? Well, there are some things in quantum physics that we just have to accept whether we find them reasonable or not. In this case, you can't really think of the particle making its way back and forth between the walls. Instead, at every instant, it occupies all the space between. The particle, when in that particular state, *is* a wave. It is spread out. To be sure, you can perform an experiment to reveal the location of the particle, but when you do, you destroy that state of motion and create a new one. Even so, the particle will never show up at points A, B, C, D, or E.

This example of a particle between walls is one-dimensional. In three dimensions, the calculation of probability is just a little more complicated. Consider, for example, the ground state of the hydrogen atom, whose wave function is shown in Figure 52. That wave function peaks at the center (at the nucleus). If it is squared to give a probability, the probability also peaks at the center. You might therefore conclude that in an experiment to pinpoint the location of the electron, you would most likely find it near the nucleus, with the steadily decreasing probability of finding it farther away. This argument needs to be refined by thinking about how much more volume of space there is farther out than near the center. The square of the wave function is the probability *per unit volume*. When you factor in the greater available volume at some distance from the nucleus than right at the nucleus, you find that there is actually a greater chance of finding the electron at some distance away than close to the nucleus. That refined probability, taking into account available volume, peaks at around 0.5×10^{-10} m.

73. How do waves prevent particles from having fixed positions?

Although a wave function has properties that can be defined at a single point—its value and its curviness—the wave itself can't be compressed to a point. A wave, by its very nature, spreads over some region of space. And it is not easily compressed to a small domain. Consider, for example, a particle confined to move back and forth between impenetrable walls (see Figure 53). As the walls are moved closer together, confining ever more tightly the range of motion of the particle, the wavelength of the confined particle gets shorter and shorter. This means, according to the de Broglie equation, that the particle's momentum, and therefore its energy, gets larger and larger. So the more the particle is squeezed within boundaries, the more energy it acquires. To squeeze it right down to where it can't move at all would take infinite energy. Paradoxically, although the particle wouldn't be going anywhere, it would have infinite kinetic energy. This would be comprehensible to a confined prisoner who, although immobilized, is exceedingly agitated.

A chunk of matter has the density it has (not to mention a texture and a color) because of quantum physics. Classically, electrons would all spiral into protons or other nuclei, radiating away energy, until all matter would be in tiny congealed clumps and all energy would be surging through the universe as electromagnetic radiation. Because of the quantum rules governing waves, electrons settle into states of motion that spread over "large" distances (that is, fractions of a nanometer). At Question 67, I discussed how the wave nature of matter dictates the size of atoms.

The transition from classical to quantum physics occurs mainly in the nanometer region. Quantum effects are evident for large molecules

FIGURE 54

Long-wavelength water waves are hardly affected by the pole. Short wavelength light is greatly affected, making it possible for you to see the pole clearly.

and other aggregations of atoms, pronounced for individual atoms and indispensable in the deeper realm of *subatomic* physics, where much of modern quantum physics is focused. Physicists want to find out what is going on in the smallest domains. The wave nature of matter makes this a challenge. It is a general rule that with a wave you can't "see" a dimension much smaller than the wavelength itself. This is what limits optical microscopy and what enables electron microscopes to reveal finer details than show up in optical microscopes, since the wavelengths of the electrons in these devices are much less than the wavelengths of visible light. You can understand this effect by thinking of a pole sticking out of the water (Figure 54). As water waves roll by it, they are hardly affected and reveal no details at all about the pole, since the wavelength of the water waves is greater than the diameter of the pole. But you can see the details of the pole readily just by looking at it—that is, by using light, whose wavelength is much less than the diameter of the pole.

Starting in the 1920s and culminating (so far) in the Large Hadron Collider in 2010, accelerators have been getting bigger and bigger, more and more energetic, and more and more expensive. (In a light moment, Enrico Fermi once extrapolated to a time in the twenty-first century when a particle accelerator would stretch around the circumference of the Earth and require most of the world's financial resources to construct.) Part of the reason for the trend is to achieve smaller and smaller wavelengths of the particles that are accelerated and thus to "see" into smaller and smaller regions of space. A proton of 7 TeV at the Large Hadron Collider, for example, has a wavelength of about 2×10^{-19} m, far smaller than the size of an atomic nucleus—in fact, about five thousand times smaller than the proton itself.

So the wave nature of matter requires more and more energy to get to smaller and smaller distances. Particles just don't like to be pinned down.

As a result, paradoxically, the largest devices of physics are used to study the smallest parts of nature. Of course, there is more to high energy than just short wavelength. The energy itself is needed to make the mass of an assortment of new, heavy particles. At the root of high-energy physics are two simple equations, De Broglie's ($\lambda = h/p$) and Einstein's ($E = mc^2$).

74. What is the uncertainty principle? Here is one statement of the uncertainty principle, offered by the twenty-six-year-old Werner Heisenberg when he introduced it in 1927: "The more precisely the position is determined, the less precisely the momentum is known in this instant, and vice versa." The principle applies not only to position and momentum but to other pairs of quantities as well, such as time and energy. In essence it is saying that the more accurately you know one thing the less accurately you can know another thing. Carried to a limit, it even means that if you knew one quantity precisely, with perfect exactitude, there is another quantity about which you would know nothing.

Werner Heisenberg (1901–1976), shown here in 1927, the year he formulated the uncertainty principle. Five years later, in 1932, he introduced the idea that the newly discovered neutron and the proton are two "states" of a single particle—the nucleon. Heisenberg headed the German atomic-bomb project in World War II. His visit to Bohr in occupied Denmark in 1941 forms the dramatic bedrock for Michael Frayn's popular play *Copenhagen*. By the time I worked at Heisenberg's institute in Göttingen in the mid-1950s, Heisenberg was a national hero frequently consulted by government figures and journalists. Photo courtesy of AIP Emilio Segrè Visual Archives, Segrè Collection.

The uncertainty principle is often said to be at the core of quantum physics, for it seems to have no counterpart in classical physics. Classically, there is no reason why you cannot know simultaneously a particle's location and its momentum, the accuracy of both measurements being limited only by technical ingenuity. The uncertainty principle provides a measure of how great the unavoidable uncertainty is, and that measure is nothing other than Planck's constant, the fundamental constant of quantum physics. (Recall my earlier discussion at Question 10: If Planck's constant were, hypothetically, reduced to zero, quantum physics would give way to classical physics, and if Planck's constant were, hypothetically, larger, quantum effects would be more pronounced. So Planck's constant sets the scale of the quantum world.) In mathematical form the uncertainty principle can be written

$$\Delta x \, \Delta p = \hbar$$

or

$$\Delta t \, \Delta E = \hbar.$$

In these equations, the Δ's stand for "uncertainty of," and the variables are position x, momentum p, time t, and energy E. On the right, Planck's constant (here divided by 2π) controls the fundamental unavoidable magnitude of the uncertainties (apart from any uncertainty that might be associated with the measurement process itself). The form of these equations helps explain Heisenberg's statement that I quoted above, for if the product of two quantities is a constant, making one of them smaller requires that the other be larger.

The time-energy form of the uncertainty principle is actually a useful tool for physicists. Some particles live so short a time that their half-lives cannot be measured directly. A short lifetime translates into a very small uncertainty in the time that the particle existed on this Earth, a small Δt. This small Δt means, in turn, a large uncertainty in the particle's energy (at least, large by quantum standards). And the energy of a particle—a particle at rest—is in its mass. So the shorter the lifetime, the greater the uncertainty in the particle's mass. Any single measurement of the particle's mass yields a particular value. Repeated measurements yield different values, and the range of values reveals an uncertainty in the mass.

As I discussed in the answer to Question 42, this mass uncertainty has been used for the Z^0 particle to calculate its lifetime, which can then be compared with theoretical expectation.

75. How does the uncertainty principle relate to the wave nature of matter? The uncertainty principle is sometimes said to be nature's way of shielding her innermost secrets, and there has been (unfortunately) no lack of volunteers seeking to apply it outside of science. Although it is indeed fundamental, the uncertainty principle can be viewed as just one more consequence of the wave nature of matter, making it a bit less mysterious, and surely less other-worldly, than it might otherwise seem.

Waves are not exclusive to the quantum domain. They are commonplace in the large-scale world of our ordinary experience as well. This means that there are forms of the uncertainty principle that exist in the classical world, too.

It is to the French mathematician Jean Baptiste Joseph Fourier that we owe an important insight about waves. He discovered that any quantity that varies from one place to another—the temperature in a room, for instance, or the density of a confined gas or the average height of buildings in different parts of Manhattan—can be expressed as a superposition of waves. The idea is illustrated in Figure 55. Part (a) of the figure shows a "pure" wave, often called a *sine wave*. It has a definite wavelength and a definite amplitude and it stretches on forever in both directions. Part (b) shows a wave that is somewhat similar in appearance to the pure wave, but is confined in space and varies in amplitude. It has about five cycles of oscillation and then dies out in both directions. According to Fourier's analysis, this truncated wave is equivalent to an endless series of pure waves added together, but not of equal amplitude. The major component is the pure wave shown in part (a). There are lesser components of waves with other wavelengths. It turns out that the predominant waves that are superposed, or mixed, to make up the wave of part (b) have wavelengths that lie within about 20 percent of the principal wavelength.

You see here already the hint of a classical uncertainty principle. The wave of part (a) has no uncertainty at all in its wavelength but a complete (infinite) uncertainty in its location, for it extends without limit in

Jean Baptiste Joseph Fourier (1768–1830). In 1789 Fourier wrote to a friend, "Yesterday was my 21st birthday, and at that age Newton and Pascal had already acquired many claims to immortality." Fourier had to wait a bit. His claim to immortality rests on work he did when he was nearly forty. Today Fourier series and Fourier transforms are in the toolkits of nearly every physicist and engineer. Fourier was an ardent supporter of Napoleon and the French revolution, and he accompanied Napoleon on his invasion of Egypt. During his three years in Egypt, Fourier helped found the Cairo Institute and gathered scientific and literary information for a later multivolume work on Egypt. Image courtesy of commons.wikimedia.org.

both directions. The wave of part (b) has no single definite wavelength. It is a superposition of waves of different wavelengths, most lying within 20 percent of the principal wavelength. So we can say that it has an uncertainty of wavelength that spans about 20 percent of the principal wavelength, and it also has an uncertainty of position, which is the range of space that it occupies.

As you might guess, if a wave is to be squeezed further to have even less uncertainty in its position, a wider range of wavelengths must be mixed (superposed). The lesser uncertainty in position goes with a greater uncertainty in wavelength. The wave in part (c) of Figure 55 has been compressed down to a single "hill." To achieve that localization (that is, that small an uncertainty of position), a very broad range of wavelengths must be superposed, giving a large uncertainty in wavelength.

The waves in Figure 55 could be classical waves, but for a moment let us suppose that they are de Broglie waves of a moving particle. The first one

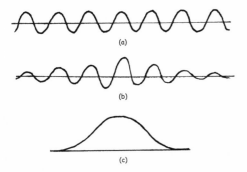

(a)

(b)

(c)

FIGURE 55

(a) A "pure" wave of a single wavelength. (b) A wave confined in space that is a mixture of different pure waves. (c) An even more confined wave with an even broader range of mixed wavelengths.

represents a particle of fixed momentum (fixed wavelength) that is spread over all space. The particle's uncertainty of momentum is zero, its uncertainty of position is infinite. We know exactly how fast it is moving and know nothing about where it is. This is a limiting case of the Heisenberg uncertainty principle. The second wave represents a particle that is partially localized (some uncertainty of position) and has also some uncertainty of momentum. The third wave has pushed the uncertainty of position to a small value, but at the cost of almost complete uncertainty of momentum.

A good example of uncertainty in the classical world can be found in the AM (amplitude modulated) radio band. There are stations broadcasting at 720 kilohertz (kHz), and 730, and 740, and so on, but none at 721 of 739 or any other intermediate frequency. Is this wasting perfectly good frequencies? No, because the act of modulating the amplitude to get the message on the airways requires mixing frequencies (or wavelengths). What is actually going out from the antenna of the 720-kHz station is a superposition of frequencies from roughly 715 to 725 kHz. You might wonder why so much frequency (and wavelength) mixing is required when the wave does not appear to be localized in space. Is it not the very epitome of unlocalized, you might ask, since it spreads throughout space? But in fact it is localized in innumerable bits and pieces. Every thousandth of a second or so, something changes in what is being broadcast, and those tiny millisecond (or even submillisecond) pieces have to be preserved. The broadcast wave is one long series of localized pieces, each of which requires a swath of superposed frequencies.

Have you ever noticed that an automatic telephone dialer takes, typically, a couple of seconds to transmit a ten-digit number? You might wonder, "Isn't this inefficient? Why not send the information in a tenth of a second or less, which should be no challenge for electronic circuits?" The answer lies in the uncertainty principle (classical version). For each "button" that is automatically pushed, two tones are transmitted with dominant frequencies that lie between about 700 Hz and roughly twice that, with "adjacent" tones being separated by about 10 percent in frequency. The shorter the duration of each pulse, the greater the range of frequencies that get mixed with the dominant frequency. If the pulse length were too short, the different tones would not be cleanly separated; the receiving circuit couldn't make out the "message." There is no danger that your fingers, dancing over the keys, will cause the tones to be garbled, but there is a real danger for autodialers. They have to be slow enough to allow for at least thirty or so oscillations of each tone before the next one starts. This assures that the frequencies will be "smeared" by three percent or less, so that the dominant frequencies will still be recognized at the receiving end. The maximum rate of autodialing is standardized at about fourteen "buttons" per second.

I have made the case that uncertainty exists in the classical as well as the quantum world, and that in both worlds localization in space requires mixing wavelengths. (Also, in both worlds, localization in time requires mixing frequencies.) Yet there remains a fundamental difference: the appearance of Planck's constant in the quantum uncertainty principle. Only in the quantum domain is wavelength related to momentum and is frequency related to energy. Thus quantum uncertainty involves material properties of mass, energy, and momentum that are not present in classical uncertainty.

76. What is superposition? Previously I have used *superposition* and *mixing* interchangeably to mean the adding together of two or more amplitudes. Think of yourself bobbing in an inner tube in a harbor as speedboats rush around in your vicinity, each making its own wave. What you feel is the superposition of all of their waves. In quantum physics, superposition has a broader meaning. Not only can wave amplitudes be superposed,

entire states of motion can be superposed.* A particle—or a system—
can be in two or more states of
motion at once. Moreover, one perfectly definite state of motion can
be a superposition of other states
of motion.†

As an example, consider an
electron with its spin directed to
the east. This is a definite state. If
you measure to find out if the spin
is directed to the east, you will find
that it is. If you measure to see if it

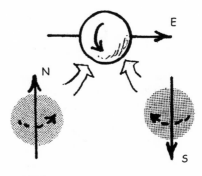

FIGURE 56
The electron's east-directed spin is
an equal mixture of north- and south-
directed spins.

is directed to the west, you will find that it is not. But suppose you are a
contrarian, and measure to see if the electron's spin is directed to the
north. There is a 50 percent chance that you will find that it is. This
is because the electron, in its definite east-pointing state, is at the same
time in a superposition of north-pointing and south-pointing states (see
Figure 56). Its east-pointing amplitude is 1.00 (making it definite) and its
west-pointing amplitude is zero. At the same time, its north-pointing
and south-pointing amplitudes are both 0.707. The reason for this number is that the square of an amplitude gives a probability, and the square
of 0.707 is 0.50, or 50 percent.

Notice here the key linkage between superposition and probability.
When states are superposed, each one of them has a certain amplitude,
and the square of any amplitude gives the probability that a measurement will reveal the system to be in that state.

A given state can be written as the superposition of other states in
many different ways. As shown in Figure 57, for instance, the electron

---

* In the mathematics of quantum mechanics, a state of motion is described by
an amplitude, so superposing states is actually a process of adding amplitudes.
† If you are familiar with vectors, these ideas won't seem so mysterious. Two or
more vectors can be added (superposed) to produce a single vector. And a single
vector can be "decomposed" into a superposition of other vectors.

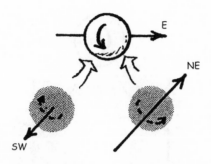

FIGURE 57

The electron's east-directed spin is an unequal mixture of northeast- and southwest-directed spins.

with its definite east-pointing spin is a superposition of a northeast-pointing state with an amplitude of 0.924 and a southwest-pointing state with an amplitude of 0.383 (respective probabilities of 85 percent and 15 percent).

This idea of superposition and probability leads to a new way to look at the meaning of a wave function. Consider, for example, the wave function of an electron in its ground state in the hydrogen atom, as depicted in Figure 52. I have already explained that the square of the wave function gives the probability of finding the electron at a particular location (actually, in a small region near that location). But an electron confined to some small region of space is itself in a state of motion, one that we might call a *localized state*. So the ground state of the hydrogen atom, as represented by the wave function in Figure 52, is the superposition of a vast number of localized states. The amplitude of each of these localized states is just the magnitude of the wave function at that location. If all of these localized states are combined (superposed) in just the right way, the result is the single state of motion that is the ground state of the hydrogen atom. Then, if you make a measurement to find out whether the electron in its hydrogen-atom ground state is actually in a certain localized state, you will find, with a certain probability, that it is, and that probability is the square of the ground-state wave function at that location. The principle is the same as that of looking to see if an electron with its spin pointing east actually has its spin pointing north.

77. Are waves necessary? Particles have wavelengths. Particles diffract and interfere. Particles have wave functions. The whole history of quantum physics since de Broglie introduced his famous equation in 1924 has been built around the wave nature of matter. Surely, you might argue, waves are an essential feature of the quantum landscape. Surely waves sit at the core of the physical world. Yet, strangely enough, the answer to the question "Are waves necessary?" is "Not really."

The wave-particle duality is too often assigned a fairy-book character, as if a particle can magically morph into a wave and back again, or be both things at once. What is that streaking by? Is it a particle? Is it a wave? Is it both? What quantum physics actually tells us is that a particle behaves as a particle when it is created or annihilated (emitted or absorbed) and as a wave in between. Measurements reveal particles. Predictions of what the results of a measurement might be use waves. The wave therefore represents a kind of possibility, or potentiality. The particle represents reality.

It was in the early 1940s, more than a dozen years after the wave-particle duality was enshrined as part of physics, that a brash young Richard Feynman came to his Princeton professor John Wheeler and said, in effect, "Who needs waves? It's all particles." What Feynman had discovered was that he could duplicate the results of standard quantum theory by postulating that a particle, from the moment of its birth until the moment of its death, did not spread as a wave of probability but instead followed all possible paths—*simultaneously*—from one point to the other. What he had done was extend the idea of superposition from waves and states to paths. He found a way to assign an amplitude to each path, and these summed amplitudes (or superposed amplitudes) gave the total amplitude for that particular fate of the particle, just the same as if a wave had been used to calculate the same thing.*

Feynman called his new approach the *path integral method*. Wheeler, who loved to coin new terms, renamed it the *sum-over-histories method*. In his autobiography,† Wheeler recalls that he was so excited by the idea that he visited Einstein at his nearby home to tell him about it and see if he could thereby shake Einstein's antipathy to the ideas of quantum

---

*Feynman's popular description of this method can be found in his charming book *QED: The Strange Theory of Light and Matter*, referenced in the footnote on page 16.
†*Geons, Black Holes, and Quantum Foam: A Life in Physics* (New York: W. W. Norton, 1998), p. 168.

physics. (Einstein, despite having been a quantum pioneer, hated the probabilistic aspects of quantum physics.) They sat down in Einstein's upstairs study, where Wheeler spent twenty minutes explaining Feynman's ideas, and then said (according to his recollection), "Professor Einstein, doesn't this new way of looking at quantum mechanics make you feel that it is completely reasonable to accept the theory?" Einstein, Wheeler recalls, was unmoved, saying only, "I still can't believe that the good Lord plays dice." (At Question 101, the last one in this book, I consider whether future developments might possibly validate Einstein's discomfort with quantum physics.)

Despite Feynman's astonishing insight, there is no reason to banish waves from quantum physics. True, waves are not *necessary*, but they are surely convenient. What is going on in the quantum world so closely mimics wave behavior that we might as well use waves to describe that world.

section XIII

Quantum Physics and Technology

78. How are particles pushed close to the speed of light? The short answer to the question is "Electrically." Charged particles can be made to go faster by electric forces.* Magnetic forces change their direction without changing their speed. Cathode ray tubes make use of these facts. Within such a tube, electrons are accelerated to high speed by electric forces, then steered to points on the screen by magnetic forces. Particle accelerators also make use of both kinds of force. In a so-called circular machine (they are not *exactly* circular), particles—most often protons—are steered around a loop by magnetic forces and pushed to higher and higher energy by pulses of electric force. In a linear accelerator, electric force drives the particles—often electrons—down a straight channel, and magnetic forces are used to help keep the particles herded into a tight beam.

An uncharged particle, such as a neutron, feels no electric or magnetic forces (except tiny ones associated with its internal magnetism), so it cannot be accelerated in the way that charged particles can. Neutrons of modest energy—up to a few million electron volts—can be secured as the result of nuclear reactions and have been used as projectiles. More

---

*They can also be slowed by electric forces, but that is not something the high-energy physicist is much interested in doing.

often, however, physicists have reason to use quite slow neutrons for experiments. Sometimes they even go to the trouble of slowing the neutrons by "thermalizing" them, which means letting them carom off atoms of cold material until they have only a fraction of 1 eV of energy (and nice large wavelengths).

Since the dawn of nuclear physics early in the twentieth century, physicists have wanted particles of higher and higher energy to probe more and more deeply into matter. Initially the available projectiles were provided by nature—first, alpha particles emitted in radioactive decay, and later, cosmic rays. The alpha particles had energies of a few million electron volts. The cosmic rays have energies ranging up to trillions and quadrillions and quintillions of electron volts—some of them much more energetic than the best that physicists have been able to produce with accelerators—but they are rather sparse, they come randomly in time and in direction, and they are a spray of many different energies. So there is reason to spend lavishly to get controlled, intense beams of defined energy.

The earliest accelerators, in the 1920s, used high voltage to achieve the electric forces needed to push protons along a straight path to energies of a few hundred thousand electron volts. In 1932, Ernest Lawrence at the University of California invented the first circular machine, later named the cyclotron. Visualize a large "pillbox," a cylinder wider than it

FIGURE 58
Essence of the cyclotron:
Charged particles circle
within the "dees," their
paths bent magnetically and
their speed increased
electrically.

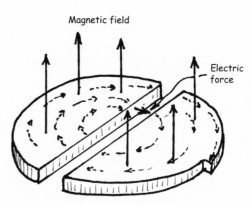

is high, cut straight across a diameter to form two hollow "dees" (Figure 58). A magnetic field causes protons (or deuterons or alpha particles) to execute half a circle within a dee. Then, as the particle crosses the gap between the two dees, it is given a little electric push. It enters the other dee at higher speed and executes a half-circle of larger diameter. Again, crossing the gap, it is nudged to higher speed and then executes a larger half-circle. This continues as it spirals out to nearly touch the outer rim of the dees. It is then "extracted" (magnetic forces again) and used as a projectile.

The special feature of the cyclotron is that the time it takes for each half-circle is the same, regardless of speed (provided the speed is much less than that of light) so the electric pulses, which must act millions of times per second, can be applied at equal time intervals (that is, at a fixed frequency). This is the key simplifying feature of the device. When the speed is greater and the effects of relativity become important, it turns out that each half-circle takes a bit longer than the preceding one, so the applied frequency of the electric pulses must decline as the particles spiral outward. At modest energy, the cyclotron can emit a steady beam. When the frequency has to be changed as the particles spiral outward, the beam is emitted in pulses. Cyclotrons have reached energies of hundreds of millions of electron volts—or roughly a thousand times more than the linear accelerators of the 1920s.

Modern circular accelerators are called *synchrotrons*. A synchrotron is like a cyclotron with its center removed. Instead of spiraling outward, the particles remain at a fixed radius, guided in a circle by magnets and accelerated by periodic pulses of electric force. As a particle gains more and more energy and its speed gets closer and closer to the speed of light, it gets harder and harder to deflect. So the magnetic force must be steadily ramped up as the particle gains energy. This synchronized electric-magnetic dance is so precisely adjusted at the Large Hadron Collider in Geneva, Switzerland that the particles execute their seventeen-mile circular path within a tube about 2 inches in diameter. Speeding around underground at a depth averaging about 100 meters, they cross the Swiss-French border more than twenty thousand times per second. The collision

Air view of the SLAC National Accelerator Laboratory. Electrons are accelerated down a 2-mile-long pipe toward the experimental area in the foreground. Positrons as well as electrons can be "stored" (briefly) in buried rings in the foreground. Photo courtesy of SLAC National Accelerator Laboratory.

of two 7-TeV protons in the Large Hadron Collider provides (or will provide—it wasn't quite there at press time) a total energy of 14 TeV, the world-record for accelerators. This is seven times more than at the runner-up accelerator, at Fermilab near Chicago, and about one hundred thousand times more than the energy of a typical cyclotron.

Linear accelerators have by no means been abandoned for circular machines. They are favored for electrons, which, in a circular machine, tend to radiate away their energy (that is, they emit photons). The current record holder for linear accelerators is the two-mile-long Stanford

Linear Accelerator in California, which pushes electrons to 50 GeV.* Electrons, because they are fundamental particle, have a certain "purity" (or lack of baggage) that makes them more attractive for some particle experiments than composite protons, with their finite size and quark components.

79. How are high-energy particles detected? The earliest particle detector was a zinc sulfide screen, viewed with a human eye. Rutherford and other early researchers learned that when energetic charged particles (initially, alpha particles) strike a screen coated with this chemical, tiny spots of light erupt where the particles land. A person with dark-adapted eyes could see the spots of light and keep track of their number and their position. It was painstaking and difficult work.

By 1911 two other detectors had been invented that were to be useful for decades to come. Rutherford's student Hans Geiger developed what is still universally called the *Geiger counter*. It consists of a thin wire running along the axis of a cylinder within which is a low-density gas, with a voltage of a few hundred volts applied between the wire and the cylinder. Because the gas in the cylinder is a good insulator, no current flows between the wire and the cylinder—at least none flows until a high-energy particle flies through the cylinder. Then a string of ions in the particle's wake makes the gas a better conductor and a brief spark jumps between the wire and the cylinder. With the right electrical circuit, the spark can trigger an audible beep.† (The high-energy particle, which can be a

---

*Whereas the early linear accelerators of the 1920s used a single electric force that acted from one end of the accelerator to the other, modern linear accelerators provide a series of electric-force pushes all along the path.

†A uranium prospector carrying a Geiger counter in the desert is hoping to detect gamma rays from radioactive minerals. Alpha and beta particles cannot penetrate the soil and the air to reach the counter. Even when sitting far from any radioactive material, a Geiger counter will click away, activated primarily by muons that streak downward from where they were created high in the atmosphere by cosmic rays.

charged particle or a gamma-ray photon, ejects electrons from molecules in the gas, transforming some of the neutral atoms into positively charged ions and negatively charged electrons. After the brief spark is quenched, the electrons quickly find their way back to the ions, and the gas is again neutral.)

The Scottish physicist C. T. R. Wilson invented the other early detector, the cloud chamber. The cloud chamber is typically a cylinder, wider than it is high, with a round glass plate on one of its circular sides so that the researcher can see into it, and with a movable diaphragm on the other circular side. Air saturated with water vapor occupies the interior of the chamber. When the diaphragm is suddenly pulled out a bit, the air cools and the vapor starts to condense into a cloud. If an ionizing particle has passed through the chamber a moment before this action, it will have left a string of ions in its wake, and the water vapor will first condense into tiny droplets on the ions, rendering the path of the particle visible for a fraction of a second before the cloud spreads throughout the chamber. If the chamber sits in a magnetic field, charged particles passing through it will leave curved tracks. Carl Anderson's discovery of the positron in 1932 made use of a cloud chamber.

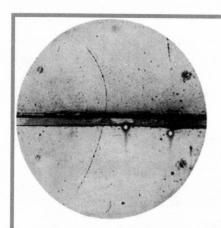

This cloud-chamber photograph by Carl Anderson, taken in 1932 and published in 1933, helped establish the existence of the positron. From the nature of the track in the chamber's magnetic field, Anderson could infer that a lightweight particle had flown through the chamber, either a negative one moving downward or a positive one moving upward. The greater curvature of the track above the central metal plate shows that the particle moved more slowly there, having lost energy as it passed upward through the plate. Photo from Carl D. Anderson, *Physical Review*, vol. 43, pp. 491–494 (March 15, 1933).

Fast-forward a hundred years from 1911. Atlas, one of six major detector arrays at the Large Hadron Collider in Geneva, consists of nested cylindrical detectors centered on the beam, extending about 150 feet along the beam and reaching out 40 feet radially from the beam. It consists of myriad different detectors and magnets and weighs as much as the iron in the Eiffel Tower. Its electronic circuitry can record about 300 million bytes of data each second. At the center of Atlas, 7-TeV protons going in one direction will collide head-on with 7-TeV protons going in the other direction, releasing 14 TeV of energy that can go into making hundreds, if not thousands, of other particles, and giving them enough kinetic energy that they can shoot out through the surrounding detector layers. The small fraction of the created particles that fly along the beam direction goes undetected. The rest fan out in all directions and trigger responses in the detector arrays.

Atlas detector at the Large Hadron Collider, partly assembled. The small tube at the center is where protons enter. Photo © CERN.

I described the Geiger counter and the cloud chamber in two paragraphs. It would take chapters to describe all that is going on in Atlas. But at their core, all of Atlas's detectors rely on the same fact about high-energy particles that makes Geiger counters and cloud chambers work—that is, particles disrupt atoms, kicking electrons free. Whether passing through tiny bits of solid silicon or through argon gas, a high-energy particle (if it is not a neutrino) leaves behind a wake of atomic "destruction" that can be pinpointed and recorded electronically. In Rutherford's day, the researcher had to squint at the scintillations on a zinc sulfide screen a few inches away. Now, thanks to the World Wide Web, the researcher can be thousands of miles away and examine detector data on a computer screen (no squinting required).

I will mention one other detector, the bubble chamber, because it was an interesting and important way station on the way from Geiger to

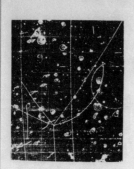 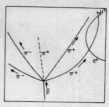

A p̄ annihilation with four charged pions and neutral pion consistent with the production of an omega meson

This bubble-chamber photograph reveals a proton-antiproton annihilation event. The antiproton flew upward in the chamber. From the point where it and a target proton met and annihilated, five pions emerged—two negative ones that flew upward to the left, two positive ones that flew upward to the right, and an unseen neutral one. Only probability governed the particular directions of the pion tracks. The photograph also shows the dance of destruction of one of the positive pions: It decayed into a muon (and unseen neutrino), and the muon then decayed into an electron (and a neutrino and an antineutrino). The small curlicues in the photograph are tracks left by lower energy electrons.

Lawrence Berkeley National Laboratory; photo and diagram courtesy of AIP Emilio Segrè Visual Archives.

Atlas. Reputedly, the bubble chamber's inventor, Donald Glaser, came up with the idea in 1952 at age twenty-five while contemplating bubbles in a stein of beer in Ann Arbor, Michigan. For his invention he was awarded a Nobel Prize in 1960. The bubble chamber contains a liquid, often hydrogen. Initially the liquid is under pressure and is held at a temperature just below its boiling point at that pressure. Then a piston is withdrawn a small distance to lower the pressure in the chamber. At this lower pressure, the temperature at which the liquid boils is lower—and indeed is lower than the temperature of the liquid at that moment (which hasn't changed appreciably). So, for an instant, the liquid is superheated, meaning that its temperature is above its boiling point although boiling has not started. If one or more ionized trails of particles exist in the liquid, they will trigger boiling, and for a few milliseconds sharply defined particle paths made up of tiny bubbles of gas will be visible in the chamber. A camera snaps a picture at just the right moment, obtaining a record of the particle paths (usually curved by magnetic force), and researchers may analyze the picture at leisure. Particle researchers used bubble chambers for more than thirty years. Some achieved gargantuan proportions, the largest having a volume of 20 cubic meters (a sphere of that volume is 11 feet across).

80. How does a laser work? In 1917, not long after completing his monumental work on general relativity, Albert Einstein turned his attention back to radiation and corpuscles of light (later to be named photons), and he discovered a process called *stimulated emission*. He didn't really discover it in the usual sense of observing it in the laboratory. Rather he ferreted it out of the quantum physics of the time in a mathematical analysis, when he was among the few who believed in particles of light. Whether you believed in photons or not, if you were a physicist of that time, you knew that an atom could absorb radiation, jumping to a higher-energy state, and could emit radiation, jumping to a lower-energy state. These two processes Einstein called *absorption* (no change of name) and *spontaneous* emission. He argued that the overall consistency of the quantum theory of radiation—and in particular, the fact that radiant energy in a cavity follows Plank's formula for the distribution of intensity among frequencies—required a third process, *stimulated* emission. Stimulated emission occurs when a photon of just the right frequency comes along and causes an atom to jump from a higher-energy to a lower-energy state, emitting a second photon that joins the first one.

For a given pair of energy states in an atom, all three processes involve photons of the same energy. In the lower-energy state, the atom can absorb a photon of given energy and jump to the higher-energy state. In the higher-energy state, it can spontaneously, with some characteristic half-life, emit a photon of the same energy and "fall" back into its lower-energy state. Or, in the newly recognized process, a photon of that same energy can strike the atom and cause it to jump to the lower-energy state sooner than it otherwise would have. One of Einstein's most remarkable conclusions was that the stimulating photon and the stimulated photon are identical not only in energy but in direction and even phase—meaning they move together and vibrate together. In modern terminology, they are coherent. Before the process of stimulated emission, there is one photon headed toward an atom. After the process there are two identical photons flying away from the atom.

In a collection of atoms, you will normally find many more atoms in a lower-energy state than in a higher-energy state. In fact, for a given temperature and for a given energy difference between the two states, a physicist can calculate how many more will be in the lower-energy state (temperature is relevant because higher temperature kicks more atoms into excited states). When a photon of the right energy enters such a collection of atoms, it will encounter more atoms in the lower-energy state and thus will be more likely to be absorbed than to stimulate emission. But if a way can be found to start with more atoms in the higher-energy state, things will be very different. Then an incoming photon of the right energy will encounter more atoms in the higher-energy state and will therefore be more likely to stimulate emission than to be absorbed. When a photon is absorbed, the number of photons goes from one to zero. When it stimulates emission, the number of photons goes from one to two. Then a chain-reaction cascade is possible in which the two go to four, the four to eight, and so on.

Physicists knew these things but for thirty years did not act to exploit stimulated emission for any practical purpose. In 1947, Willis Lamb and Robert Retherford at Columbia University harnessed the process in a particular experiment in hydrogen, and in 1954, Charles Townes, also at Columbia, invented the maser (an acronym for *microwave amplification by stimulated emission of radiation*). But let me jump right to the invention of the laser (light amplification by stimulated emission of radiation) in 1958, which resulted eventually in a torrent of practical application. In that year, Townes and Arthur Schawlow, a Bell Labs researcher, published a paper suggesting how to make a powerful coherent beam of light using stimulated emission. In the same year, they applied for a patent on the process.

I can best explain the Townes-Schawlow idea by describing the first working laser, constructed by Theodore Maiman at Hughes Research Laboratories in California in 1960 (the same year in which the Townes-Schawlow patent was awarded—and in which Maiman himself applied for a patent). It consisted of a ruby crystal in the shape of a rod, with parallel mirrors at the ends of the rod. Surrounding the rod was a tube similar to a neon tube that could emit a flash of green light. When a circuit triggered the flash, chromium atoms in the crystal were propelled to an excited state, call it C, from which they quickly (in about ten millionths of a second) decayed to a lower-energy state B that was still 1.8 eV in energy above the ground state A (see Figure 59). That lower-energy, but still excited, state is what is called *metastable*—it lives a "long" time, about a thousandth of a second—and more atoms end up in state B than in state A. The stage is set for a cascade of stimulated emission. Let's say that one of

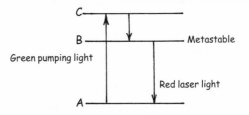

FIGURE 59

Schematic energy-level diagram for chromium atoms in a ruby crystal. Outside green light excites the atoms to state C, from where they quickly decay to the metastable state B, resulting in more atoms in state B than in state A and allowing for a cascade of stimulated emission.

the atoms in state B undergoes spontaneous emission and sends a photon hurtling toward a mirror at one end of the rod. That photon might trigger another just like itself on the way, or it might make it to the mirror, reflect, and have another chance to stimulate emission. And every time a new photon is added to the cascade by stimulated emission, it joins in the process of stimulating still more photons. In an instant, the laser is filled with photons of red light, all of the same wavelength and all traveling back and forth along the rod's axis. That is a laser pulse. When one of the mirrors is made semitransparent, some of that laser light escapes in an intense narrow beam. The energy in the laser beam is much less than the original energy supplied by the flash of green light, but it is very useful energy because it is such a narrow, intense beam.

The ruby laser is not yet a museum piece. It still finds uses—for instance, in holography and cosmetic surgery—but it has been joined by countless other lasers, large and small, with wavelengths running from the infrared to the ultraviolet and beyond.* The one feature that all lasers have in common is "population inversion"—that is, establishing more atoms in a higher-energy state than in a lower-energy state. Once that is achieved, in whatever way and with the help of reflecting mirrors, the laser cascade occurs. Some modern lasers, like the original ruby laser, are pulsed lasers, in which, after each laser pulse, the atoms must again be "pumped" into a higher-energy state. Others, such as the ubiquitous helium-neon laser, operate continuously, with population inversion maintained steadily.

81. How do electrons behave in a metal? A metal is an element (or alloy) that readily conducts electricity (and heat). It is called a *conductor*

---

* The imaginative, charismatic Hungarian-born physicist Edward Teller once advocated using X-ray lasers as weapons to shoot down enemy missiles. In his scheme, an atomic bomb was to provide the excitation (a green flash tube wouldn't come close). X-ray laser action was actually demonstrated in an atomic-bomb test, but nothing came of the idea of turning such a laser into a weapon. Scientists at Livermore National Lab subsequently developed laboratory-scale X-ray lasers, using extremely intense pulses of light to provide the excitation.

and is to be distinguished from an insulator, which is a material that does *not* readily conduct electricity and heat. The disparity between the conductivities of conductors and insulators is quite astonishing, more than a billion billionfold in some cases. No wonder that electric current finds it easier to travel 100 meters down a copper wire than 1 millimeter sidewise through the insulation encasing the wire.

To clarify the difference between conductors and insulators, and to highlight how electrons behave in such materials, let me compare two elements, sulfur (an excellent insulator) and copper (an excellent conductor). The sixteen electrons in a single atom of sulfur spread themselves (that is, spread their wave functions) over a dimension of about a tenth of a nanometer and arrange themselves in overlapping shells. The outermost shell contains six electrons (two short of being full). Removing an electron from an isolated sulfur atom—that is, ionizing the atom—takes an energy of 10.4 eV (the energy of a "hard" ultraviolet photon). When sulfur atoms are clustered together in a solid, the electrons remain tethered to their individual atoms. The outermost electrons, the so-called valence electrons, have no more freedom to move away from their host atoms than if the atoms were isolated. Actually, between the energy of the tethered electrons and the energy of escape (ionization) there is a region of energy called the *conduction band*. If an electron gets into that band, it can move freely through the solid material. But to get to that band the electron has to be kicked with an energy that is not much less than what it would take to remove it entirely from the material. Neither thermal agitation nor any ordinary applied voltage can supply enough energy to the electrons to elevate them into the conduction band, so they stay put in what is called the *valence band*. Of course, in physics as in life, one should never say "Never." A very few electrons in sulfur can make it to the conduction band, so the conductivity of this element is not strictly zero. But it is 10^{23} times smaller than the conductivity of copper.

An isolated copper atom is somewhat larger than a sulfur atom (about 50 percent larger). Twenty-eight of its twenty-nine electrons occupy closed shells. Its twenty-ninth electron, occupying its own larger shell, is the valence electron. To remove an electron from a copper atom—to ionize it—takes 7.7 eV, about three-quarters as much energy as to ionize a sulfur atom. That difference is all-important. When copper atoms are clustered into a solid metal, the conduction band is not only more accessible than in sulfur, it actually overlaps with the valence band. That means that the valence electrons in copper, with no extra energy added at all, find themselves in the conduction band and are free to move through the material. The valence electrons become conduction electrons. This behavior is typical of all metals. Copper is almost the champion conductor. Only silver beats it out, and not by much. There are some thirty-two metallic elements whose conductivities are within a factor of ten of copper's.

There is one more thing to say about electrons in metals. Conduction electrons are not immune to the exclusion principle. Just as no two electrons within a single atom can occupy the same state of motion, no two conduction electrons in a metal may share a state of motion. The difference is that because of the vast difference in the physical space available for the electrons' motion, there is a corresponding vast difference in the spacing of the energy levels—well separated in an atom, pressed infinitesimally close together for conduction electrons. For that reason, we refer to *bands* for electron motion in a solid rather than *states* or *energy levels,* the terms used for atoms. On the other hand, the number of electrons in an atom is small, whereas the number of electrons in the conduction band is enormous. As a result, as I mentioned at Question 22, the total energy spread of a band from bottom to top is several electron volts, comparable to the spacing between individual energy states in an atom.

82. What is a semiconductor? A semiconductor, as its name implies, is a "not-quite," or "almost," conductor. The most ubiquitous semiconductor, used in just about every electronic circuit in the world, is the element silicon (also a major constituent of sand and glass). In very round numbers, the conductivity of silicon is about a trillion times less than that of copper and about a trillion times more than that of sulfur. Silicon is element number 14. It has ten electrons in filled shells and four valence electrons in its outer shell, making that shell both half full and half empty (four less electrons in that shell is found at the noble gas neon, four more at the noble gas argon). The "band gap" in silicon—that is, the energy needed to elevate an electron from the valence band to the conduction band—is 1.1 eV, much less than the gap in sulfur and other insulators, yet still large enough to inhibit all but a tiny fraction of the valence electrons from making it into the conduction band. So silicon can conduct electricity, but not readily.

There is a way to turn silicon into a better conductor—and in fact to control just how well it will conduct electricity. That is to add a small amount of "impurity," typically of another element. The impurity is added while the silicon is a liquid (at high temperature). Then, when it cools and its atoms settle into a crystalline arrangement, some of the places where in pure silicon a silicon atom would reside are now occupied by one of the impurity atoms. This process is called *doping*, and it takes two forms. In one form, the added atoms have five valence electrons (in contrast to silicon's four). Each such atom will contribute an extra electron to the crystal, which is free to roam through the material. Overall, the crystal will be electrically neutral, but that neutrality is achieved by having positive ions (of the impurity) at fixed locations and negative electrons that migrate readily from place to place in the material. Such a "doped" semiconductor is called *n type* (*n* for *negative*) and the impurity atoms are called *donors* (since they donate electrons). Phosphorus and arsenic are among the elements whose atoms can serve as donors. By controlling the amount of the impurity, the conductivity is controlled.

Here is an analogy to a semiconductor and its n-type alteration. A ballroom is packed with tables that seat four, and there are four people at every table. Once in a great while one person will get up and wander around, but for the most part everyone remains seated. That corresponds to pure silicon, with its low conductivity. Then a few parties of five are ushered in and asked to replace parties of four at some of the tables. Only four of each such new party can take seats. The fifth person wanders about, greatly increasing the number of wandering guests. That corresponds to an n-type semiconductor, with its increased (but still not large) conductivity.

The other way to dope a semiconductor is with an element with three valence electrons in its outer shell. Then at the site of each impurity atom, there is one less electron than there would be in the pure semiconductor. This absence of an electron is called a *hole*, and it takes on a life of its own. An electron in an atom adjacent to an impurity atom can jump over, providing four electrons at the impurity site but leaving a "hole" in the neighboring atom. If the electron jumps to the right, the hole moves to the left. The hole acts like a positively charged particle, and it can wander around in the doped semiconductor. A semiconductor modified in this way is called *p type* (*p* for *positive*) and the impurity atoms are called *acceptors* (they accept electrons from their neighbors). Boron and aluminum are among the elements whose atoms serve as acceptors.

To return to the ballroom and its four-person tables, think now of a few parties of three ushered in and asked to replace parties of four. Wherever the new party sits there will be an empty chair. Some friendly person from a neighboring party of four can move over to occupy the empty chair, and in doing so leaves an empty chair behind. Others can move from table to table to occupy empty chairs, giving the impression that a few empty chairs are moving around in the ballroom. Even though people are only shifting from seat to seat, there is more mobility in the room than when it was entirely occupied by parties of four. This evidently corresponds to the p-type semiconductor, with its increased conductivity supplied by holes.

83. What is a p-n junction? Why is it a diode? P-type and n-type semiconductors become interesting, significant, and exceedingly useful when they are joined in what is called a *p-n junction*. Let me start with the ballroom analogy and then go back to the real thing. Suppose that the ballroom is divided down the middle into halves, with the west half having a certain number of parties of five so that some guests are wandering around that half without a place to sit, and the east half having a certain number of parties of three so that some empty chairs crop up as guests move one after the other to sit in the available chairs. Near the boundary, some of the extra guests in the west half will notice the available chairs in the east half and move across to occupy them. Soon there will be a region extending on both sides of the dividing line in which there are neither standing guests nor empty chairs. After a time, the wander-

ing guests in the west half will decide it is not worth their while to go all the way across the middle region to get to an empty chair that is now some distance away. So an equilibrium is established in which there are three regions—west, with an excess of guests; middle, with neither extra guests nor empty chairs; and east, with some empty chairs.

In a p-n junction, the middle region near the point of contact of the n-type and p-type semiconductors is called a *depletion zone,* a zone depleted in both electrons and holes. On a human-sized scale it is thin, but on an atomic scale it is thick—up to tens of thousands or hundreds of thousands of atoms across. What makes the p-n junction reach an equilibrium (or, to use the language of the analogy, what makes an electron in the n-type region "decide it is not worth its while" to cross over to the p-type region) is a voltage difference established across the depletion zone. That puts a stop to the flow of electrons by repelling them to their own side.

Now the physicist may apply an external voltage to the p-n junction, in addition to that provided by the junction itself. If that external voltage is in the same direction (or has the same sign) as the internal voltage, it reinforces the effect of opposing electron flow. No current results. If the external voltage is opposite to the internal voltage and larger in magnitude, it overpowers the retarding effect of the internal voltage. It not only allows electrons to flow from the n-type region to the p-type region, it encourages them to do so. A current results. So the p-n junction is a diode, a circuit element that allows current to flow in one direction and inhibits or stops current in the other direction.*

To return (for the last time) to the ballroom analogy, imagine that the ballroom, with its three regions, is located on a ship. When the ship rolls, elevating the east side, the extra guests on the west side will be even less likely to cross over to the remaining empty chairs on the east side. No

---

*When I studied circuits in the pre-solid-state era, a diode was a vacuum tube containing a heated cathode that emitted electrons and a small cold plate that could attract or repel these electrons. When a positive voltage was applied to the plate, it attracted electrons, and current flowed through the tube. When a negative voltage was applied, current was stopped. (Heated cathodes are still found in the cathode ray tubes of some computers and television sets.)

"current" of people results. When the ship rolls to the opposite side, elevating the west side, the extra guests on that side will gain extra incentive to move across (downhill) to the east side. A "current" of people results. (Pretty soon, the west side of the ballroom would run out of extra standing people and the current would stop. The p-n junction, connected to a circuit, doesn't suffer this fate. The external applied voltage provides a continuing, nonstop supply of electrons to replenish those on the n-type side.)

84. What are some uses of diodes? The p-n junction diode can be used as a one-way valve to allow current to flow in one direction and prevent it from flowing in the other direction. But it has found a variety of other uses as well, some of great practical value, all of them related to the fact that in the p-n junction electrons are perched on one side of the depletion zone, ready to jump to the other side if given the right "motivation." Here I discuss several such applications.

In the light-emitting diode, commonly abbreviated LED, an applied voltage sends current through the diode, and some of the electrons literally leap across the divide, emitting photons as they do so. The energy gap between the two sides of the diode can be regulated by the choice of material (which need not be silicon) and by the kind and amount of impurities. This means, in turn, that the color of the light can be regulated, since the color of light is determined by its photon energy. A gap of 1.5 eV or less results in invisible infrared light; a gap of 4.0 eV or more, in invisible ultraviolet light. Gap energies between 1.5 and 4.0 eV can be used to achieve any color of visible light from red to violet. As long ago as the 1960s, red LEDs made an appearance in the displays of the first handheld calculators. Since then, scientists and engineers have learned how to make LEDs not only of all colors but also of greater intensity. They are now seen in flashlights, traffic signals, automobile brake lights, and, most prominently, in large video displays in cities and at sports stadiums. Much more widespread use can be expected in the future, including automobile headlights and home lighting. Like other solid-state devices, LEDs are longlasting, and they are much more energy-efficient than incandescent bulbs.

In the right circumstances, the light-emitting diode can be turned into a laser diode. It is not quite as simple as "just adding mirrors," but almost. With electrons on one side of the diode awaiting a trigger to jump to the other side, one has, effectively, population inversion, with more higher-energy than lower-energy states filled. The current that triggers the electrons to jump and to emit photons also keeps replenishing the electrons, thereby maintaining the population inversion. If a pair of mirrors encloses the diode, light emitted in one direction can be reflected and can stimulate more photons to be emitted. These photons can, in turn, stimulate more photons, building up the intensity in one direction and producing a laser beam. The essential features of every laser, population inversion and stimulated emission, are encapsulated in the tiny diode. At supermarket checkout counters, laser diodes are replacing older gas-tube lasers. And scarcely any lecturer can get by without a laser diode pointer.

The two applications of solid-state diodes that I have just discussed involve the emission of light. Other applications make use of the reverse process, absorption of light to create electric current. When a photon is absorbed by a p-n junction, it can propel an electron upward (upward in energy) from the p-type side to the n-type side of the junction. In effect, it creates an electron and a hole and sends them to their respective sides. The added energy provides a voltage that can spawn a current. The net result is the exact opposite of what happens in an LED. Instead of current creating light, light creates a current. The simplest application is a light-activated switch—which might better be called a *dark-activated switch*. Usually, in that application, a light beam maintains a steady current. When something blocks the light beam, the light-activated current stops, and that throws a switch.

For many reasons, not least of which is a sustainable future for humankind, the most important application of p-n junction diode may be the solar cell, or photovoltaic cell. In some applications, such as powering a handheld calculator or meeting the power needs of astronauts in orbit, the cell's efficiency—that is, what fraction of the incident light energy is converted to electrical energy—is not of paramount concern. But for large-scale application, in which arrays of solar cells replace fossil-fuel power plants, cost and efficiency are critical parameters. Much of the research on solar cells over the last half-century has gone into making them more efficient. Electrons and holes will do their part to save the planet.

85. What is a transistor? The transistor has been called the most important invention of the twentieth century. It is not difficult to comprehend in principle, but its invention in 1947 (announced in 1948) by John Bardeen, Walter Brattain, and William Shockley at Bell Labs in New Jersey required a deep understanding of the quantum physics of solids, an understanding that emerged only after quantum principles had been applied to atoms and molecules. (Bardeen, Brattain, and Shockley shared the Nobel Prize in physics in 1956 for their invention. Bardeen went on to win another Nobel in 1972 for his work on superconductivity.) Here I present an overview of a basic form of the transistor.

First imagine a piece of silicon doped with arsenic or some other impurity to make it an n-type semiconductor (Figure 60a). It has a certain conductivity as determined by the amount of the impurity. Now slice it in half and insert a thin piece of p-type semiconductor in the space between the two halves (Figure 60b). In this way you have created two p-n junctions back-to-back. But the important thing is that you have disrupted the conductivity of the original semiconductor and put yourself in a position of control. You can control what current flows through the device by what voltage you apply to the thin inserted piece. If a voltage difference is applied to the two ends of the original piece—say negative on the bottom, positive on the top—electrons will be inclined to flow from bottom to top. If you apply a positive voltage to the central thin piece, that will attract electrons and increase the flow. Some of the electrons will be diverted out the side through the central piece, but most will continue to the top side

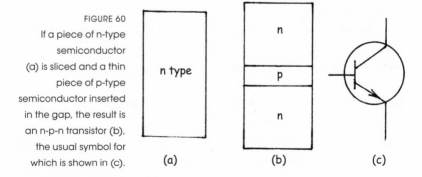

FIGURE 60
If a piece of n-type semiconductor (a) is sliced and a thin piece of p-type semiconductor inserted in the gap, the result is an n-p-n transistor (b), the usual symbol for which is shown in (c).

n type

n

p

n

(a)   (b)   (c)

of the original semiconductor. If you apply a negative voltage to the central piece, it will repel electrons and greatly decrease the flow of current.

So the transistor is an amplifier. A small variable voltage applied to the central piece will result in a large variable current through the device. The transistor can be a switch as well. Changing the sign of the voltage on the central piece can switch the main current on and off. The transistor is, in effect, a "triode."* It has three elements and three connecting wires to the outside world. For the transistor pictured in Figure 60, the bottom is called the *emitter* (it emits electrons), the top is called the *collector* (it collects electrons), and the central piece is called the *base*. The transistor in the figure is called, for obvious reasons, an *n-p-n transistor*. Equivalent performance can be supplied by a p-n-p transistor.

Modern transistors are not separate circuit elements, they are imbedded in a silicon crystal; they are connected not to the outside world by wires, but to neighboring imbedded devices by tiny bits of conducting material; and they can be considerably more complex than the simplified Figure 60 indicates. But the marvelous technical developments that have occurred since 1948 are a story for another book.

---

* In the pre-solid-state days of electronics circuits, a triode was a vacuum tube containing three elements: a heated cathode, which emitted electrons; a positively charged plate called the *anode*, which collected electrons; and between them a wire mesh called the *grid*, whose voltage controlled how much current flowed from cathode to anode (actually, conventional current direction is opposite, positive to negative). The transistor's three elements exactly mimic the elements of the old triode—except that the transistor is smaller (*much* smaller), dissipates less energy, and has no cathode to burn out.

section XIV

Quantum Physics at Every Scale

86. Why do black holes evaporate? Black holes were named (by John Wheeler, in 1968), taken seriously, explored theoretically, and finally identified in nature, all without invoking quantum physics. They seemed to be wholly classical objects. Actually, Wheeler at one point hoped that quantum physics would save the world from black holes. He said that he did not want to believe in these strange entities in which matter and energy are crushed to a point (a "singularity"). He hoped that some attribute of quantum physics at the subatomic scale could put a stop to the collapse and prevent it from being total. Perhaps, he thought, a heavy star at the end of its life might collapse to a very small entity, but not all the way to a black hole. However, all of his efforts to use quantum physics to find an escape hatch failed. He concluded, finally, that the black hole is the inevitable fate of a large enough mass of "cold, dark matter" (relatively cold, that is, after nuclear processes have run their course).

Quantum physics reentered the black-hole story in a rather roundabout way. When black holes first became objects of serious study, theoretical calculation showed them to be very simple entities—identified by mass, electric charge, angular momentum, and nothing else. Whatever properties the matter that collapsed to form the black hole (or that was later sucked into it) may have possessed—such as lepton number, baryon number, quark flavor, and so on—were lost. This is usually described as a loss of information, and even today physicists are divided about whether information is lost in a black hole or just put in storage for later possible recovery. Wheeler expressed the apparent simplicity of black holes by saying that "black holes have no hair."* He was visualizing a room full of people who would be easier to distinguish from each other if they had hair than if they were bald. Yet Wheeler was troubled by "hairless" black holes. Their simplicity seemed to imply that they had little or no entropy—entropy being a measure of disorder. It was puzzling that a black hole that might have been formed from a complex, disordered collection of matter should not inherit the large entropy of the matter that went into it.

As Wheeler tells it in his autobiography,[†] he sat down with his graduate student Jacob Bekenstein in 1972 and said something like (I am paraphrasing) "When I put a cup of hot tea next to a glass of iced tea and let them come to a common temperature, I feel like a criminal, for I have contributed to an increase in the entropy of the universe, an increase that will echo to the end of time; there is no way to erase or undo it. But if I drop the hot tea and the cold tea into a black hole that happens by, I have committed no crime. I have not contributed to an increase of entropy of the universe. I am spared punishment." Bekenstein took this half-joking remark as a challenge. He disappeared for several months, then returned to Wheeler's office to announce that Wheeler would not be spared. A black hole, he said, does have entropy, and its entropy is measured by the area of its "event horizon." (The event horizon is the sphere surrounding a black hole that separates the regions from which escape is possible and from which escape is not possible.) Dropping the tea into the black hole increases its mass slightly, thereby increasing the area of its horizon—and, according to Bekenstein, increasing its entropy. Stephen Hawking had already proved, two years earlier, that the area of a black hole's horizon can never decrease. That is consistent with Bekenstein's conclusion, for entropy, too, can never decrease (that is one way to state the second law of thermodynamics).

---

* Richard Feynman chided Wheeler for being naughty.
† *Geons, Black Holes, and Quantum Foam: A Life in Physics* (New York: W. W. Norton), p. 314.

At first, apparently, leading black-hole researchers, including Hawking, clung to the idea that a black hole has little or no entropy and so they doubted Bekenstein's assertion. (Wheeler, as he recalls it, said, "Your idea is so crazy that it might just be right. Go ahead and publish it.") Yet Hawking pondered Bekenstein's claim and began to see that it had merit. Indeed it inspired him to reach a most remarkable conclusion about black holes. In 1974 he introduced the idea that black holes are not wholly cut off from the outside world after all. They can radiate. Not only do they have entropy, they also, as a corollary, have temperature, and, accordingly, they radiate away energy (and mass), just as the Sun radiates because of its surface temperature. Hawking's work made it possible to calculate the temperature of a black hole, which depends only on its mass, and in turn to calculate the intensity of what has come to be called *Hawking radiation,* or sometimes *Bekenstein-Hawking radiation.* A black hole with a mass of a few Suns has a temperature less than a ten-millionth of a degree above absolute zero, and radiates at such a leisurely rate that its mass will not noticeably decrease over billions of years. Smaller black holes have higher temperatures and greater rates of radiation. A sub-microscopic black hole, if such exists, might have such a high temperature that it could radiate away its remaining mass in fractions of a second and make for a quick astronomical event. Astronomers have looked, so far in vain, for evidence of the final moments of tiny black holes.

Now quantum physics reenters the tale. Physicists accepted the validity of Hawking's theory and the reality of Hawking radiation. But they still had to ask "What is the *mechanism?* How can a black hole emit energy if, supposedly, nothing can escape from it?" The only credible resolution of the apparent paradox has been provided by quantum physics. When virtual particles, the uncertainty principle, and the wave nature of matter are all taken into account, the horizon ceases to be an entirely sharp boundary. It gets a little fuzzy. Then, just as with tunneling, what is impossible classically becomes possible quantum-mechanically. Picture a virtual pair of particles coming into existence near the horizon (Figure 61). In normal empty space, such virtual pairs are coming into and immediately going out of existence all the time (see Question 7). But when such a pair is created next to a black-hole horizon, the enormous pull of gravity may be enough to separate them permanently. One is sucked into the black hole, the other is expelled into the cosmos. The black hole loses mass in the process—very, very slowly at first, eventually at a great rate.

What are these particles that are separated and expelled? They are almost all photons, at first. The black hole's temperature is so low that only zero-mass particles will take part. So the Hawking radiation is electromagnetic radiation, at least until the black hole becomes tiny and its temperature becomes large. Then neutrinos can enter the virtual particle dance, and eventually other, more massive particles as well.

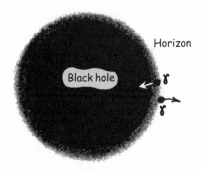

FIGURE 61

Near the slightly fuzzy horizon of a black hole, a pair of virtual particles, instead of quickly recombining, can be separated forever, with one being "radiated" away.

87. How does quantum physics operate in the center of the Sun?

The temperature at the center of the Sun is about 15 million degrees Celsius—hot, but, without the aid of quantum physics, not hot enough. At that temperature, protons are flying about with an average kinetic energy of some 2,000 eV. In a head-on encounter, two protons with that energy can get no closer to each other than several hundred fermis because of the electric repulsion between them. A fermi is 10^{-15} m, about the size of a single proton, so these protons, even in a head-on encounter, are far from touching and far from being able to fuse and release nuclear energy. In the mix, some protons have kinetic energies greater than the average. This lets them get closer together. Fewer than one in a trillion have an energy as great as thirty times the average. Even the rare proton with that much energy can push up against electric repulsion only to within about 12 fermis of another proton.

According to classical physics, then, even at the very high temperature in the Sun's center, protons can't get close enough together to fuse and release energy (of course, without such energy release, the Sun's center wouldn't *be* that hot). That quantum tunneling might play a role in the Sun and allow nuclei to get together and fuse was suggested as early as 1929 by the Welsh physicist Robert Atkinson (then thirty-one) and the German physicist Fritz Houtermans (then twenty-six). This was only a year after tunneling had first been proposed as an explanation for alpha decay (see Questions 35 and 70). Atkinson was later honored for his work on stellar nucleosynthesis, the idea, now well established, that all heavy elements result from fusion reaction in stars. Houtermans, an avowed Communist, had a rough time. He took a job in the Soviet Union in the 1930s but came under suspicion and was imprisoned. Following the Hitler-Stalin pact of 1939, he was turned over to Nazi Germany and imprisoned there, but released in time to join the German atomic bomb project. Following World War II, he was among the German scientists interned at Farm Hall in England.*

Atkinson and Houtermans imagined a process in the Sun that involved heavy nuclei acting as catalysts for the fusing of light nuclei. Eight years later, in 1937, George Gamow, one of the original scientists to propose the tunneling phenomenon, and Edward Teller had the idea of a fusion cycle that involved carbon nuclei as catalysts. Gamow, originally from Russia, and Teller, an émigré from Hungary, were both then at George Washington University in Washington, D.C., and were busy planning a 1938 conference that was to focus on energy in stars. They attracted Hans Bethe (a German émigré then at Cornell University) to the conference, where Bethe, already a leading nuclear physicist, became captivated with nuclear reactions in stars. In 1939 Bethe published a paper on energy generation in the Sun (and other stars), which took account of quantum tunneling and provided, for the first time, a detailed theory for understanding fusion in stars. For this work he was awarded the Nobel Prize in physics in 1967. Bethe later credited Teller for setting him on this path, saying "Edward is responsible for my fame."[†]

---

* The Farm Hall story is told by Jeremy Bernstein in *Hitler's Uranium Club: The Secret Recordings at Farm Hall* (New York: Springer, 2010).

† Later, in the 1950s, Bethe and Teller had very divergent views on the development of thermonuclear weapons and gave contrary advice to the government, but continued to respect each other's scientific abilities and to be at least superficially cordial.

Tunneling, an odd little feature of quantum physics, has turned out to be the reason stars shine.

88. What is superconductivity? Perpetual motion is commonplace in the small-scale world. Electrons in atoms never slow down. Nucleons never get tired. And in the cosmos, friction is so small that motion—of planets and stars and galaxies—is very nearly perpetual. But in the every-day world around us, perpetual motion is practically unknown. Don't try to patent a perpetual-motion machine. You won't succeed. No wonder that Aristotle thought that a force is necessary for anything to move. In our ordinary experience, anything that isn't pushed or pulled comes to rest. And electric current, even in good conductors, dies out if not kept going with an outside voltage.

Superconductivity and superfluidity are the exceptions to the rule against perpetual motion in the human-sized world. In 1911, Heike Kamerlingh Onnes, a Dutch physicist at Leiden University, discovered that the element mercury, when cooled to the temperature of liquid helium, which is 4 degrees above absolute zero—where mercury, like everything else *except* helium, is solid—conducts electricity without resistance. Not merely small resistance, but literally *zero* resistance. Such a material is called a *superconductor*. To see what zero resistance means, visualize a superconductor fashioned into a doughnut shape with a current circling in it. That current, without any further urging from outside, will continue circling indefinitely. It is as if an electron circling without resistance within a single atom has had its orbit inflated a billionfold.

Over the years, scientists discovered various features of superconductors, such as the fact that magnetic fields don't exist within them; and they found new materials that were superconducting at ever higher "transition temperatures," about 18 kelvins (that is, 18 degrees above absolute zero) in the 1950s, then a jump to 30 kelvins in the 1980s, and now above 100 kelvins. Why these so-called "high-temperature" superconductors (it's all relative) perform as they do remains a mystery. But a theory put forward in 1957 by three American physicists—John Bardeen, Leon Cooper, and John Schrieffer—has found solid support as a description of what is going on in the lower-temperature superconductors. Let me try to give at least a rough idea of the so-called BCS theory (which, in its full rendition, is highly mathematical).

For a good conductor—for instance, copper or aluminum—we say that electrons move through the material relatively freely. That is what sets a conductor apart from an insulator. But even in the best of ordinary conductors, the electrons do encounter some friction (that is, some resistance) as they occasionally exchange energy with positive ions in the crystal lattice through which they flow. In these encounters, the electrons lose some energy and the ions gain energy, which is then dissipated through the material as heat. A conductor carrying electricity gets warm. As the temperature of a conductor is decreased, the thermal agitation of its ions decreases, the energy losses to the ions by passing electrons decreases, and the resistance gets smaller but usually does not reach zero. (Most superconductors, paradoxically, are not good conductors at room temperature.)

Underlying the BCS theory is the discovery (a theoretical discovery) by Leon Cooper that in certain materials and at low enough temperature, electrons can form bound pairs. Cooper's then colleagues at the University of Illinois, David Pines and John Bardeen, had already concluded that electrons, despite their electrical repulsion, could attract one another weakly in certain solids. Cooper added the insight that the weak attraction could be enough to form pairs (*Cooper pairs*, as they are now called) capable of moving through the material as separate entities. Here is the mechanism of the attraction: An electron traveling through the material attracts nearby positive ions toward itself, pinching them together, so to speak. Then a following electron, seeing a slight concentration of positive charge ahead, is drawn toward it, effectively being attracted by the electron ahead. This has been likened to a bowling ball on a mattress, which creates a trough and "attracts" another bowling ball toward itself, even though it is the mattress, not the first bowling ball, that is doing the attracting. If the deformation of the mattress is great enough, the apparent attraction of the two bowling balls could persist even if they carry like charges that tend to push them apart.

Because a Cooper pair is made of two fermions, it behaves as a boson, which means that many pairs can nestle together in the same state of motion. The exclusion principle does not act on them. To separate the pairs or pull one apart takes energy—not much energy, to be sure, but at extremely low temperature there is not enough thermal energy to disentangle the pairs, which then move without resistance through the material. (One of the oddest things about the pairs is that the separation of the two electrons making up a pair is greater than the average separation of one pair from another. It's roughly like a crowd of semitrailer trucks on a multilane highway, where the length of one truck is greater than the average separation of the trucks—except that for the Cooper pairs there is only one lane.) What I am describing here is only a pale reflection of the full theory, but it captures one essential point: Cooper pairs sneak unimpeded through the crystal lattice because the lattice has insufficient thermal energy to destroy them.

Any current flowing in a loop generates a magnetic field that threads through the loop. That is how electromagnets are constructed. If the loop is made of a superconductor, a new quantum effect shows itself. The magnetic flux—which is a combined measure of field strength and area—is quantized. Like angular momentum, it can take on only values that are integral multiples of a smallest value. The explanation of this effect bears a startling resemblance to de Broglie's first idea on why atomic orbits in atoms are quantized. In a superconducting loop, the wave function of the current has to complete an integral number of cycles in going once around the loop so that it positively reinforces itself and maintains the supercurrent. The quantum of magnetic flux is small, but not too small to measure in our human-sized world. If you form a circle about an inch in diameter with your thumb and middle finger, the flux of Earth's magnetic field through that circle will be about 10 million quantum units.

Needless to say, superconductors are a boon to technology. At both Fermilab and the Large Hadron Collider, the magnets that deflect protons are made using superconducting wire. Comparing these magnets with conventional magnets, the expense and trouble of maintaining the superconductors at a temperature about 2 degrees above absolute zero is more than offset by the savings in power consumption. It seems likely that superconductors will also find use in magnetically levitated vehicles of the future.

89. What is superfluidity? In 1937, Pyotr Kapitsa,* working in Moscow, and, almost simultaneously, John Allen and Donald Misener, working in Toronto, discovered that liquid helium, cooled below 2 kelvins, exhibits the remarkable property of flow without friction (without *viscosity*, to use the technical term). Such flow is called *superfluidity*. Researchers later discovered the same phenomenon in the isotope helium-3, but at a much lower temperature, below a thousandth of a kelvin. Superfluidity and superconductivity have in common the facts that they are frictionless motion and that they bring a quantum effect into the large-scale world. They also resemble each other in that the flow is propagated by bosons. Helium-3 superfluidity most nearly resembles superconductivity because it involves bosons formed from pairs of fermions. In superconductivity, the fermions are electrons. In helium-3 superfluidity, the fermions are whole helium-3 atoms. (Counting protons, neutrons, and electrons, a helium-3 atom contains five fermions, making it also a fermion.)

The mechanism of superfluidity in helium-4 is somewhat different in that helium-4 atoms are already, individually, bosons without pairing. At temperatures below 2 kelvins, the helium-4 atoms can coalesce, overlapping in the lowest available energy state. This clustering, which is really Bose-Einstein condensation, inhibits the transfer of energy to mechanical vibrations (to sound waves) and thereby puts a stop to dissipation—that is, to the transfer of energy from ordered to disordered motion, which is what friction and viscosity are all about.

One fascinating practical application of superfluidity was in the satellite-borne experiment called *Gravity Probe B*. This experiment, which sought evidence of extremely subtle twists in spacetime near Earth (its data are still being analyzed), used a gyroscope that had to be insulated

---

*Kapitsa had left the Soviet Union to work at Cambridge University in England, but on a visit back to his home country in 1934, he was detained by Stalin and forced to remain—free to work but not free to leave. Ernest Rutherford arranged for Kapitsa's lab equipment to be shipped to him in Moscow so that he could continue his work there.

from any outside influence that could tend to tilt it or slow it. This was achieved by submerging the gyroscope in superfluid helium at a temperature of 2 kelvins. At the same time, superconductivity played a role. A superconducting coating on the gyroscope provided a signal showing its exact orientation.

90. What is a Josephson junction? In 1962, as a twenty-two-year-old graduate student at Cambridge University, Brian Josephson got interested in the possibilities associated with the tunneling of electrons from one superconductor to another through a thin insulating layer that separated them (an arrangement that later came to be called a *Josephson junction*). He made two remarkable discoveries. The first was that even with zero voltage across the insulating layer, a current flows, demonstrating that it is a true tunneling current rather than a weak conventional current driven by voltage. Moreover, the current can flow in either direction and with any magnitude up to some maximum that is dependent on the relationship of the wave functions of the Cooper pairs in the two superconductors (what is called their *phase difference*). This was clearly a large-scale quantum effect, for it depended on the behavior of quantum wave functions and involved tunneling through a classically insurmountable barrier.

Josephson's second discovery was spookier still. He found that if a constant voltage is applied across the junction, an alternating current (AC) is generated. (With conventional materials, a constant voltage could only drive a constant, direct current, so-called DC). Moreover, Josephson predicted the frequency of this AC to be equal to the applied voltage multiplied by a simple constant involving the quantum unit of charge e and Planck's constant h. (Wherever h appears, you can be sure that quantum physics is at work.)

It need hardly be said that for these insights, soon borne out by experiment, Josephson was granted a doctoral degree. Before long (in 1973) they also brought him a Nobel Prize in physics.

When two Josephson junctions are connected in parallel, as shown in Figure 62, still another astonishing effect makes its appearance. If the magnetic field in the "hole" of this arrangement is zero, then, as with a single junction, current can flow, half through one junction, half through the other (assuming that the junctions are identical). But if a magnetic field—more exactly, a magnetic flux, which is field times area—is threaded through the hole, the quantization of flux through a superconducting loop (discussed at the previous question) enters the picture. Any effort to install a flux there that is *not* an integral multiple of the basic quantum unit of flux will be automatically thwarted. A small extra current will flow around the loop to create an extra small magnetic field, just enough to bring the total flux to a quantized value. The superconductor will not "allow" a flux that is not a multiple of the flux quantum. The effect of this extra small current, which could be, say, left to right at the top of the loop and right to left at the bottom, is to alter the phases of the wave functions along the upper and lower paths and so alter the total current that flows through the pair of junctions. In effect, the junctions then act just like a pair of slits that can lead to constructive or destructive interference—in this case, interference between the currents following the upper and lower paths. This interference can be controlled by the magnetic flux that is applied—or is attempted to be applied— through the loop.

This arrangement of two Josephson junctions is known as a *SQUID* (*superconducting quantum interference device*). One of its practical uses is to measure magnetic fields to a level of precision never possible before. For every change in flux of just 1 quantum unit, the current through the device passes through a cycle of constructive and destructive interference, easily observed. As I mentioned above, 1 quantum unit of flux is

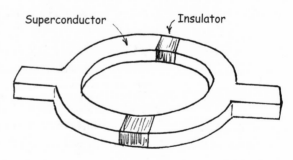

FIGURE 62
A "SQUID," the result of forming a loop with two Josephson junctions. It is capable of measuring magnetic fields to exquisite accuracy.

about 10 million times less than the flux of Earth's field through a 1-inch-diameter loop.

91. What is a quantum dot? In grade school I had my first introduction to the idea of an atom by being told that if you cut a piece of material in half, then cut each of those halves in half, and keep cutting the smaller and smaller pieces into still smaller pieces, eventually you could go no further because you would have reached an uncuttable single atom. I didn't think to ask what, if anything, changes when you get to a piece that is forty atoms across, or twenty, or ten. A piece of material that small is called a *quantum dot*, and it does have interesting properties distinct from the properties of both bulk matter and single atoms. If the dot is made, say, of a doped semiconductor, it still has valence-band and conduction-band electrons, even if it is as small as ten atoms across. But those bands, instead of having the nearly continuous range of energies they have in a large chunk of matter, have identifiably discrete energies, much like the individual quantized energies in single atoms. The larger the number of atoms in the quantum dot, the more closely spaced are the energies in the bands. So, as the diameter of the dot expands from one atom to five to ten to a hundred to a million, the spacing of the electron energies shrinks from a few electron volts to a fraction of an electron volt to, finally, such a small value that the range of energies can be treated as continuous.

Something else changes as the size of the dot gets larger or smaller. The energy gap between the conduction electrons and the valence electrons changes. As the size of the dot gets smaller, the band gap gets larger. This means that photons emitted when electrons jump from the conduction band to the valence band are more energetic for smaller dots. If a certain dot emits red light, a smaller one made of the same material emits green light (photons of greater energy), and a still smaller one emits blue light (still more energetic photons). This opens the possibility of using quantum dots (lots of them!) as light sources of chosen color.

That quantum dots have the properties they do is not so surprising and could have been forecast many years ago. What they represent is not so much a triumph of theoretical understanding as a triumph of technology. The ability to make quantum dots of controllable sizes and in vast numbers is a tour de force in the modern field of micromanipulation known as *nanotechnology*.

> *The questions in this section of the book have ranged, so far, from black holes in the cosmos down to collections of a few atoms. In the balance of the section we dive deeper.*

92. What is a quark-gluon plasma? Heat dissociates* matter—that is, pulls it apart. Consider, for example, a chunk of solid nitrogen at, say, 50 kelvins. If it is heated to a little more than 63 kelvins, it will melt. Its molecules will lose their rigid hold on each other and will be able to swim about in the liquid nitrogen. Now heat the liquid to 77.5 kelvins and it will boil. The molecules will gain more freedom as they carom around almost freely in the gas. The gas consists of molecules of N_2, two nitrogen atoms bound together into a simple molecule. As the gas is heated further, to white-hot and beyond, the atoms remain tied together in N_2 molecules, but when the temperature surpasses 20,000 kelvins or so (from here on, my temperature numbers are quite approximate), the atoms begin to come unglued, and by the time 50,000 kelvins is reached, the gas is made up almost entirely of nitrogen atoms, not molecules—actually, ions (charged atoms), not just neutral atoms, for at this temperature electrons are being dislodged from the atoms by the turbulence of thermal motion. As the temperature continues to climb, first one, then two, and finally all seven electrons are freed from each atom. At a temperature of around 5 million kelvins, what was once our solid chunk of nitrogen has become a plasma of bare nitrogen nuclei and electrons. (Note, incidentally, that what started out as a single "particle," a nitrogen molecule, has become sixteen particles—two nuclei and fourteen electrons.)

---

*To a chemist, *dissociate* means breaking apart a molecule into its atomic constituents. I am using the term in a broader sense to mean pulling matter apart to reveal any underlying parts.

As the temperature is nudged ever higher, this plasma will continue to exist, until a temperature around 50 billion kelvins is reached. Then the intense jostling of nucleus against nucleus begins to break apart the nuclei, spilling out their proton and neutron components. At around 200 billion kelvins, this process is nearly complete, and, for every original molecule, there are now forty-two particles bouncing around in the proton-neutron-electron plasma (fourteen protons, fourteen neutrons, and fourteen electrons).

We can infer that this is not the end of the line. Although the electrons are fundamental particles with no known prospect of being broken apart, the protons and neutrons are composite particles containing quarks and gluons. What temperature will it take to split open the nucleons? Theoretical calculations suggest a temperature of a little more than a trillion kelvins will do that job, creating a so-called quark-gluon plasma, a soup of quarks, gluons, and electrons. Can such a temperature be reached? For perspective, consider that the temperature at the center of the Sun—about as hot a place as we know anything about—is only about one one-hundred-thousandth of this quark-gluon temperature. Yet there is one way to reach such a temperature, at least for an instant. That is to slam one high-energy nucleus against another in an accelerator. An accelerator designed for exactly this purpose exists at Brookhaven National Laboratory on Long Island, New York. This accelerator is the Relativistic Heavy Ion Collider, or RHIC (pronounced "*rick*"). It can strip gold atoms of all seventy-nine of their electrons and accelerate the nuclei to an energy of 20 TeV, then collide two such energetic nuclei together to make a total of 40 TeV available in the fireball of a collision. (This is even more energy than is available with protons in CERN's Large Hadron Collider—a difference made possible by the fact that the gold nuclei accelerated at RHIC have seventy-nine times as much charge, and therefore seventy-nine times a much susceptibility to electric force, as the protons accelerated at the Large Hadron Collider.)

Data have been accumulating at RHIC since 2000, with no certain conclusions yet on the detailed properties of the quark-gluon plasma. It is rightly called a new state of matter—supplementing the already-known solid, liquid, gas, and conventional plasma. What is learned about it will shed light on the earliest moments of the universe. According to calculations, a temperature above a trillion kelvins existed for only about a millionth of a second after the Big Bang. Then the universe "cooled" to a point where the quarks assembled themselves into nucleons. Since that moment, it is possible that no quark-gluon plasma ever again existed anywhere until being brought to life at RHIC.*

93. What is the Planck length? What is quantum foam? In 1899, even before advancing the idea of quantized energy exchange, Max Planck in Germany had come upon a constant that he called b (and the next year called h). He considered it fundamental because it entered into a formula for cavity radiation, a phenomenon independent of any particular material. He then found that if he combined his new constant b with two other established fundamental constants of nature, Newton's gravitational constant G and the speed of light c, he could obtain a quantity with the dimension of length. It was, to be sure, an exceedingly small length, a few times 10^{-35} m, but it appealed to Planck because, unlike, say, the meter, its definition did not depend on any human choice of a comparison length (the meter was originally defined as one ten-millionth of the distance from the Equator to a Pole). Planck's calculated length unit was, for many years, a curiosity without any apparent significance since it was so many orders of magnitude smaller than any length that appeared in physics experiments or theory. That changed in 1955 when John Wheeler was exploring where general relativity—Einstein's theory of gravitation—and quantum physics might make contact.

As Wheeler put it in his autobiography, at the dimension of a nucleus or single proton, where "we see the expected frenzied dance of particles, the quantum fluctuations that give such exuberant vigor to the world of the very small," the arena for all of this, space and time (spacetime), remains

---

*AS of 2010, researchers at RHIC had achieved a temperature of 4 trillion kelvins and were seeing some evidence that the quark-gluon plasma was behaving more like a liquid than a gas. Researchers at CERN in Geneva have also studied quark-gluon plasmas and are expected eventually to harness the Large Hadron Collider in the hunt for still hotter bits of matter.

FIGURE 63
Quantum foam. A
fanciful view of what
space and time
might look like at the
Planck scale.

"glassy smooth." There should come a point, Wheeler reasoned, where spacetime joins the quantum dance. He credits discussions with his student Charles Misner for helping him reach the conclusion that that point is at the dimension of Planck's calculated length, which Wheeler renamed the *Planck length* (and which some physicists call the *Planck-Wheeler length*). If there is a Planck length, there is also a Planck time, the time it takes light to travel one Planck length. That time is about 10^{-43} second.

In this Planck realm, where spacetime and quantum physics embrace, the "glassy smoothness" of spacetime is expected to give way to fluctuations of geometry, including writhing, multiply connected regions, which Wheeler christened "quantum foam." Figure 63 shows a fanciful rendition of this condition.

The dimension where quantum foam is expected to show itself is small beyond our ability to comprehend. The number of Planck lengths that, if lined up, would stretch across one proton is the same as the number of protons that, lined up, would stretch from Philadelphia to New York (remember, one hundred thousand protons reach only across one atom). Yet it is in this incredibly small Planck domain where important new physics of the future may be found (which is not to rule out the possibility that lots of new physics awaits us between the nuclear scale and the Planck scale).

section XV

Frontiers and Puzzles

94. Why are physicists in love with the number 137? Most of the quantities we deal with come with a unit: 6 feet, 3 miles, 2 days, 5 pounds, 40 watts. Change the unit and you change the number. Six feet is 2 yards; 3 miles is 4.83 km; 2 days is 48 hours; and so on. But some numbers are "pure," or dimensionless. The number of sides on a die is six, no matter in what units you measure the size or the weight of the die. And the ratio of two quantities with the same unit is also dimensionless. If it is fourteen miles to your aunt's house and two miles to your favorite restaurant, the ratio of those two distances is 7. You can measure the distances in meters or rods or feet or furlongs. The ratio will still be 7.

Most of the quantities used in the quantum world also come with units: atomic mass units or electron volts or nanometers or picoseconds, among others. Some significant numbers are dimensionless. The mass of a muon, for example, is 206.769 times the mass of an electron, no matter what unit is used to measure mass. And there is one particularly intriguing combination of quantities that is dimensionless. Start with Planck's constant divided by 2π ($\hbar$, "h-bar"). Multiply it by the speed of light (c) and divide by the square of the charge of an electron (e). The result, $\hbar c/e^2$, is a dimensionless number equal to 137.036. No matter what set of units are used for $\hbar$, c, and e, the result is the same. Back in the 1920s, when these quantities were less precisely known, some physicists thought this number might turn out to be *exactly* 137. No such luck.

Why is 137.036 significant, not just intriguing? Because its inverse, $e^2/\hbar c = 1/137.036$, or roughly 0.007, is a measure of the strength of the electromagnetic interaction. It is what is called a *coupling constant*, coupling charged particles to photons, the quantum carriers of electromagnetic energy. Because that number is small, the coupling is rather weak, and physicists are able to calculate processes of emission and absorption of light with relative ease. This inverse constant, 1/137.036, is also known as the *fine-structure constant*. Because of relativity, energy levels in atoms are "split"—that is, separated—depending, for instance, on the relative orientation of spin and orbital angular momenta. This splitting, which is small relative to the separation of energy levels first explored by Niels Bohr, is called *fine structure*, and it depends on the fine-structure constant. If the fine-structure constant were larger, the splitting would be larger. If the fine-structure constant were as large as 1, there would be no fine structure, only gross structure, in atomic energy levels. (A lot of other things would be different, too, and we probably wouldn't be here to discuss it.)

Why the numerical value 137? No one knows. It's a question physicists hope will be answered one day. One of those fascinated by this dimensionless number was John Wheeler, whom I have mentioned in other contexts in this book. One day in the early 1990s, he came upon a new sidewalk being laid along Washington Road in Princeton. He went up to the foreman, turned on his combination of charm and persuasiveness, and somehow convinced the foreman that history and scholarship would be well served if he were allowed to scratch a number in the still-wet concrete. There, near the corner of Washington Road and Ivy Lane, remains to this day 137.036 carved in the sidewalk. The photo shows Wheeler pointing to his handiwork in 1994.

All of the particles we know anything about are either electrically charged or neutral (uncharged). So they are coupled to the electromagnetic field (photons) either weakly or not at all. But another class of particles is theoretically possible—particles carrying magnetic charge, or "pole strength." These so-called monopoles were investigated by Paul Dirac in England in 1931. They would be direct sources of magnetism. Electrically charged particles, by contrast, are indirect sources of magne-

John Wheeler (1911–2008). There was a bit of an imp in Wheeler. He loved this carving. Photo by the author.

tism. They generate magnetic effects by moving, not just by "being there." Dirac's striking discovery was that if monopoles exist, they would be coupled to photons much more strongly than electrons or protons are. The magnetic coupling constant would be 137, not 1/137. Over the years, there have been many experimental searches for monopoles (I participated in one of them in 1963), and none has been found.

Strongly interacting poles create no theoretical problems in classical physics. They would just be particles that happen to push and pull on each other nearly twenty thousand times more strongly than electrons do. But they do create problems in quantum physics, where new things happen when interactions are strong. There is no known lower limit in how weak an interaction might be, but there could be an upper limit on how strong one can be. I have long harbored the hope that some theorist will figure out such an upper limit for magnetic interaction and thus provide a *reason* for the number 137. Perhaps the electric coupling constant 1/137.036 is as weak as it is because the magnetic coupling constant 137.036 is as large as it is.

95. What is entanglement? Books could be written on entanglement (and have been*). The basic idea is simple, but the demand on our credulity is great. Entanglement is a particular kind of superposition—one in which the superposed states may spread over large distances, far beyond the normal range of quantum effects. Any quantum state can be expressed as a superposition of two or more other states. As I discussed at Question 76, for instance, an electron with its spin pointing north is, at the same time, in a superposition of states with its spin pointing east and west. And an electron in the ground state of a hydrogen atom is, at the same time, in a superposition of countless states, each having the electron localized at a particular place. These kinds of superposition, although mind-stretching, are not as brain-straining as the kind of superposition known as entanglement.

Consider, for example, a neutral pion at rest that at some moment decays into a pair of photons. The initial pion has zero charge, zero momentum, and zero angular momentum (spin). Because these are conserved quantities, their values are also zero for the products of the decay. The two photons have zero charge. That takes care of charge conservation. They fly apart back-to-back with equal momentum. That takes care of momentum conservation, since the vector sum of two oppositely directed equal vectors is zero. And their spin directions must be opposite to preserve zero total spin. But the two photons that fly apart are, until observed, a single system—a superposition of oppositely directed momenta

---

*See, for example, Louisa Gilder, *The Age of Entanglement: When Quantum Physics Was Reborn* (New York: Knopf, 2008), and Amir Aczel, *Entanglement* (New York: Penguin, 2003).

and oppositely directed spin. Suppose that one photon flies off to the left, the other one off to the right. The spin of the left-going photon could be directed to the left, in which case the spin of the right-going photon must be directed to the right. A shorthand for that state might be [L, R]. Or, if the left-going photon's spin is directed to the right, the right-going photon's spin must be directed to the left. That state can be called [R, L]. The actual state that is created in the decay process is an equal mixture (superposition) of these two possibilities (for those chosen directions of flight). That state can be written [L, R] − [R, L].* This means that if you measure the direction of the left-going photon's spin, you have an equal chance of finding it directed to the left or to the right.

This is where Einstein's "spooky action at a distance" comes in. You could measure the spin of that left-going photon a meter or a mile or a light-year from where it was created. At the moment you make the measurement, you can conclude what the spin direction of the other photon must be—two meters or two miles or two light-years away. Establishing with certainty the spin direction of one photon determines the spin direction of the other photon at that instant, even though, up until the measurement is made, the spin directions of both photons are uncertain and unknown. All of this because the two photons, until the moment of measurement, constitute a single quantum system, not two separate quantum systems.

In 1935, two years after arriving in the United States as an émigré from Germany, Albert Einstein joined with two colleagues at Princeton's Institute for Advanced Study, Boris Podolsky and Nathan Rosen, to publish one of the most famous papers ever written about quantum physics, even though the paper's purpose was only to question quantum physics, not to advance it. One might almost say that the paper's purpose was to question the sanity of anyone who could accept an argument like the one I just gave about one part of a system snapping to attention when a measurement was made on a remote and seemingly completely disconnected other part of the system. EPR—as these authors have, forever

---

*There is a subtle reason for the minus sign, which is beyond the scope of this discussion. The essential point is that the two possibilities have equal probability.

after, been called—described the situation they were addressing in this way: Two systems come together, interact, and then separate. According to quantum physics, the two systems, by virtue of their interaction, have become *entangled* (this word was introduced by Erwin Schrödinger soon after the EPR paper) and that, until some measurement is made to disentangle them, they are in fact—no matter how spread out—a *single* system with superposed amplitudes and relative probabilities no different than if they continued to occupy the same confined space. EPR argued that the idea that one system could react instantaneously to a measurement made on another system meters or miles or light-years away made no sense, and so must be wrong. By stretching the idea of superposition to what they considered an absurd limit, EPR thought they had found a flaw in quantum physics—or at least a flaw in the way that physicists described the theory.

The idea that a measurement in one place might reveal information about something in another place is not in itself so strange. Suppose that Joe and his sister Mary go traveling, one to Tokyo, the other to Paris, but you don't know who went where. If you meet Joe in Tokyo, you know at once that Mary is in Paris. Nothing odd about that. But if some quantum physicist tells you that what is in Tokyo is some superposition, or mixture, of Joe and Mary, and that it is your encounter on the street, your "measurement," that establishes Joe to be there and reveals Mary to be in Paris—and moreover that it is equally likely that you could meet Mary in Tokyo and thus learn that Joe is in Paris—you might understandably say, "That's absurd." Absurd it may be. And brain-straining it may be. But in the quantum realm it is true.

Entanglement is not the only thing in quantum physics that goes against common sense. But it is one of the most vivid—and was, for some of the twentieth century's outstanding scientists, one of the most troubling. Experiments in various parts of the world over the past few decades have confirmed the predictions of quantum physics concerning the behavior of entangled systems. A measurement at one place does indeed determine an outcome at another place. A simple way to say it is that so far all predictions of quantum physics have, when tested, been found to be correct, whether or not they conform to our "common sense," whether or not they seem to defy logic.

96. What is Bell's inequality? In the 1960s, the Irish theoretical physicist John Bell, then in his thirties, was working at CERN in Geneva. His main job was to work with experimental physicists in planning and interpreting experiments at the lab's accelerators, but he found time to devote to his first love, fundamental quantum theory. In his heart, he didn't *like* quantum theory. He shared with Einstein, Podolsky, and Rosen a distaste for the idea of fundamental probability and, above all, for the idea of entanglement. Yet in his head he believed (correctly, as it has turned out so far) that quantum theory would always accurately predict the results of experiments.

For nearly thirty years after its 1935 publication, the EPR paper remained famous but not particularly influential. No one could think of a way to decide experimentally whether a pair of particles shooting from a common origin were entangled or not. Then in 1964 Bell published a

John Bell (1928–1990). Raised in a poor Irish Protestant family in Belfast, Bell was the first person in his family to pursue education even as far as high school. As a teenager, he became disillusioned with his first love, philosophy, and decided that physics was "the next-best thing." Graduating from high school at sixteen, he had to find a job for a year before he was old enough to enter the university, and the job he found was as a lab assistant at the local Queen's University, where, he said, "I did the first year of my college physics when I was cleaning out the lab and setting out the wires for the students." He became, in time, one of the deepest thinkers of the twentieth century on quantum physics. (The quotations are from Jeremy Bernstein, *Quantum Profiles* [Princeton, N.J.: Princeton University Press, 1991]). Photo © CERN.

short paper that was destined to become not only famous but influential. In it, Bell showed that indeed the reality of entanglement could be tested experimentally.

The idea can be illustrated with a simple example—for which I turn myself into an experimental physicist. In the lab, as shown in Figure 64, I arrange for an electron and a positron—particles of spin ½— to be created and to fly apart back-to-back with zero total angular momentum (that is, with canceling spins). At some distance apart, I set up detectors to sense spin "up" or "down." First, as shown in part (a) of the figure, I orient the detectors identically, with their up-down direc-

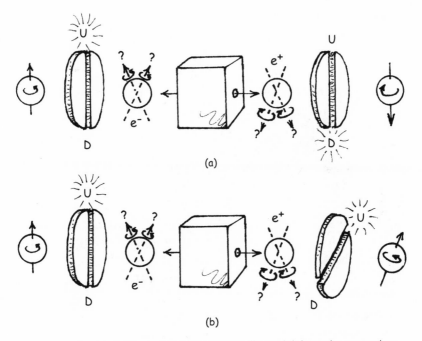

FIGURE 64 Back-to-back electron and positron with zero total angular momentum (spin). (a) Detectors are set to record spin along an "up-down" axis. If the left detector flashes "up," the right detector is sure to flash "down." (b) The right detector is rotated through a small angle. If the left detector flashes "up," the right detector may (with small probability) also flash "up."

tions matching the vertical direction in the lab. According to quantum mechanics, the pair of particles are in an entangled state that can be expressed as a superposition of spins up-down and down-up. That means that there is a 50 percent probability that detector 1 will register "up" while detector 2 is registering "down," and a 50 percent probability of the reverse. There is zero probability of the detectors registering up-up or down-down. The up-up correlation is said to be zero. This just means that one "up" reading is never matched by another "up" reading by the two detectors.

If you choose to side with EPR and believe that the particles had quite definite spins ever since they separated, you need not be troubled by these results. You are free to assume that if the spin is angled upward (pointing anywhere above the horizontal plane), it will be flipped by a detector into an "up" orientation, and if it is angled downward, it will be flipped by a detector into a "down" orientation. Because you expect a random orientation of the spins in all directions, you won't be surprised that the up-down and down-up results each occur half the time and the up-up and down-down results don't occur.

Now I will change the experimental setup by rotating detector 2 through a small angle, as shown in part (b) of the figure. With this arrangement, from the perspectives of both quantum mechanics (entangled spins) and EPR adherents (definite spins in advance of detection), occasional up-up readings and down-down readings should occur. The quantum physicist's explanation is that when the electron is found by detector 1 to have spin up, the positron must have spin vertically downward, which is a superposition of the new "down" and the new "up" directions on the right (a lot of the new "down," a little of the new "up"). The EPR adherent's explanation is that for the occasional electron whose spin is angled slightly upward and so triggers an "up" reading at detector 1, the matching positron will have a spin angled slightly downward, but by less than detector 2's rotation angle, so it records an "up" reading also. By either interpretation, there is now an up-up correlation that is small but no longer zero.

Enter John Bell. He was able to demonstrate that according to the EPR view of spins fixed in advance of measurement, doubling the small angle of detector 2 can at most double the up-up correlations. This is a remarkable result, since it doesn't depend on what law might be invoked to "instruct" spins how to flip when encountering a detector. It is an *inequality** because it can be written: The up-up correlation for twice the angle is less than or equal to twice the up-up correlation for the original angle. By contrast, quantum mechanics has something specific to say about how the up-up correlation changes when the angle doubles: For most angles the correlation more than doubles, and for quite small angles it quadruples. Thus quantum mechanics is said to "violate" the Bell inequality, because it predicts as much as a quadrupling of correlation whereas the EPR concept of spins fixed in advance predicts at most a doubling of the correlation.

Why the quantum quadrupling? Basically, it is because quantum mechanics deals with amplitudes, and probabilities are proportional to the squares of amplitudes. So when an amplitude doubles, as it does in this case, the probability quadruples.

Testing whether Bell's inequalities (as noted in the footnote, there are more than one) are valid is not easy. Such experiments require sophisticated setups with exquisite measurements of time. The first such experiments were conducted in the 1970s and they continue up to the present day—in Vienna, Paris, Geneva, and elsewhere. All, to date, have found that the predictions of quantum mechanics for entangled states are borne out and that the inequalities are violated—as Bell expected they would be. The electron and positron in Figure 64 do not let go of each other as they fly apart.

97. What is a qubit? What is quantum computing? A bit[†] is an elemental piece of *classical* information: a yes or no, a zero or one, an up or

---

* This is only one example of an inequality, and it is not the original example presented by Bell. Since his 1965 paper, other related inequalities have been derived, collectively known as Bell *inequalities*.

† The term *bit* was coined by the Princeton statistician John Tukey. Tukey was a delightful person who also knew a lot of physics. I enjoyed participating in a folk-dance group that he helped found in Princeton.

down, an off or on.* A qubit (pronounced "CUE-bit") is an elemental piece of *quantum* information. Yet these two kinds of "bits" are dramatically different. A classical bit is analogous to your options in driving on a north-south street. You can drive north or you can drive south. Just two possibilities. Nothing in between. A qubit is analogous to your options in driving on a large paved lot. All directions are available to you. But it's worse. It's really analogous to being able to drive in any two opposite directions at once on the lot. The spin of an electron provides an example. It can be measured to point up or point down. Just two possibilities. Then it acts like a classical bit. But *before* it is measured, it can exist in a superposition of up and down directions, with any relative mixture. It could be 50 percent up and 50 percent down, or 31 percent up and 69 percent down, or 91 percent up and 9 percent down.

This remarkable feature of the qubit—its ability to exist as a superposition of different states—has led, over the last few decades, to accelerating interest in quantum computing (an exciting possibility that is still a long way from practical realization). At the heart of ordinary computers are "logic gates," circuit elements that process bits. For example, a logic gate could provide an output bit that depends on what two bits are fed into it, or it could function as an on-off switch depending on what single bit is fed into it. A quantum logic gate has the potential to process *both*

---

*Nowadays we speak more often of bytes. A byte is a string of 8 bits. One bit can distinguish just two possibilities. One byte can distinguish 256 possibilities, enough to encode upper- and lowercase letters, numerals, and assorted symbols. In 1951 and 1952 I used a computer that occupied two rooms and had a total storage capacity of about 2 kilobytes (2×10^3 bytes). Those rooms would now easily accommodate two hundred computers, each with a terabyte (10^{12} bytes) of electronic storage. For every single byte of stored information in that long-ago computer there would now be 100 billion bytes. The total number of bytes presently stored in the world is mind-numbingly large and growing fast. It may have surpassed the total number of brain neurons in all of humankind around 2010. The number of bytes of data stored in a typical handheld device today greatly exceeds the total number of bytes stored worldwide sixty years ago.

up and down, or zero and one, bits at the same time. This capability can lead to much more than a twofold increase in processing power if multiple qubits are superposed. Two qubits can be mixed in four ways, ten in a thousand ways, twenty in a million ways. In principle, a quantum logic gate could process all these possibilities at once—*provided* it can do so without disturbing the system, for a disturbance is equivalent to a measurement that extracts one of multiple possibilities. A quantum computing theorist must ponder how a superposed system can survive a processer, emerging on the other side to be analyzed later.

Eventually, of course, information must be extracted from the quantum computer. The superposed states can't just roam around undisturbed within the computer forever. When that information is extracted, indeed the superposition is disturbed and a single outcome among all the possibilities is realized. This need not invalidate the usefulness of the quantum computer, for often only a single answer is wanted. Questions such as "Is this humongous number a prime?" or "Will it snow in Vail tomorrow?" can be answered with a yes or no, although a vast amount of computing might be required to arrive at the answer.

A classical bit is more than a zero or one written on a piece of paper. To be useful in practice, it has to be realized in some physical device. It could, for example, be a pulse of light running down an optical fiber or a speck of magnetized material on a hard disk. Similarly, if quantum computers are to become real, qubits must become physical "things." At this early stage of quantum computing, there are myriad possibilities for what those things might be. Among the suggestions are Josephson junctions (see Question 90), quantum dots (Question 91), and even individual atomic nuclei. No doubt the field will narrow, just as it did years ago for classical bits, which were once represented by holes punched in cards, spots on the faces of cathode rays tubes, and pulses of sound in gas-filled tubes.

98. What is the Higgs particle? Why is it important? Satyendra Nath Bose and Enrico Fermi possess the distinction of having classes of particles named after them (bosons and fermions). But only Peter Higgs, an emeritus professor of physics at Edinburgh University, shares his name with an individual fundamental particle*—a particle whose discovery was announced by CERN scientists in Geneva, Switzerland, on July 4, 2012 (not a holiday in Switzerland).

In the early 1960s, particle theorists were united in predicting that if their theories were correct, there should exist in nature a boson of zero charge, zero spin, and zero mass. No such particle was known. If it existed and interacted with other matter as it was expected to do, it could hardly have escaped detection. In 1964 Higgs saw a way out of the conundrum of the expected but unobserved fundamental particle. He discovered within the mathematics of relativistic particle theory (field theory) a loophole, so to speak. This loophole, which has come to be called the *Higgs mechanism*, permits, or even demands, that this particle have non-zero mass—possibly even a quite large mass (by particle standards). Thus was born the Higgs particle, or, as it is affectionately called, just the Higgs. Like the originally envisioned particle, the Higgs is a boson with no charge and no spin,[†] but very far from massless. With a mass of 125 GeV (more than 130 times the mass of a proton), it is heavier than any other known particle except the top quark.

Soon new duties were assigned to the Higgs. By the early 1970s, theorists had concluded that the field[‡] underlying the Higgs particle might account not only for the mass of the Higgs particle itself, but for the masses of many particles. The Higgs field, filling all of space, is expected to provide a kind of viscosity that impedes particles and gives them mass. This is obviously a highly oversimplified statement, since for all the fundamental particles that *do* have mass, no two have the *same* mass, and certainly there is no theory that predicts what their masses should be. Yet at least one oddity in particular is likely to find its explanation in the Higgs. That is the enormous disparity in the masses of the force car-

---

* Higgs has written and spoken good-humoredly of his "life as a boson." I don't know how he feels about having his namesake particle sometimes referred to as the "God particle."

† Because it has no spin, it is necessarily a boson.

‡ In this book I have focused on particles, not fields. But they go together. For example, the photon is the particle associated with the electromagnetic field and the graviton is the particle associated with the gravitational field.

riers for the electromagnetic and weak interactions. As discussed at Question 16, theory has united these two interactions of very different strength into a single electroweak interaction, but with one force carrier, the photon, having no mass, and two force carriers, the W and Z particles, having enormous mass. Theorists expect that the Higgs field may provide not only the "viscosity" that gives the W and Z particles their masses, but also the "cement" that unites these particles and the massless photon.

What does the Higgs do, once it is created? In its lifetime of probably less than 10^{-23} seconds, it moves no farther than the diameter of a small nucleus, then decays into other particles. A single Higgs could decay into, say, a bottom quark and bottom antiquark, among many possibilities. These particles in turn decay into others, and so on until some final entities such as photons and electrons fly far enough to be detected. The two CERN teams that discovered the Higgs used two massive detectors, one being Atlas (see the photo on p. 213). Their formidable task was to figure out the mass of some initial particle that triggered all the final debris, a little like deciding from which Rocky Mountain spring a drop of water in the lower Mississippi River emanated.*

99. What is string theory? Like most of the other topics treated in this last section of the book, string theory has called forth entire books aimed to explain the subject to the nonspecialist.† Here I touch on just some of the main features of this mind-stretching theory.

Aristotle said that nature abhors a vacuum. Modern physicists abhor singularities—that is, things or events that occupy no space and extend over no time. Yet quantum physics does contain singularities. The electron and other fundamental particles are assumed to have no spatial ex-

---

* Fermilab came close. After the Tevatron shut down for budgetary reasons in the fall of 2011, researchers there continued to analyze its mountains of data. On July 2, 2012 (not a holiday in Illinois), they announced tantalizing but not quite solid evidence of a new particle with a mass between 115 and 135 GeV.

† Two good ones, both by Brian Greene (and both with lengthy subtitles not reproduced here), are *The Elegant Universe* (New York: Vintage Books, 2000) and *The Fabric of the Cosmos* (New York: Vintage Books, 2005).

tension, and interactions between particles are assumed to happen at points in space and moments in time.* A hardy band of contemporary theorists have set out to do something about this. Instead of merely abhorring singularities, they are trying to get rid of them. According to their extension of quantum physics (and gravity), called *string theory*, particles do not exist at points. They are instead tiny vibrating strings, either loops or lines with two ends. The strings are unimaginably small and their rate of vibration unimaginably large. They exist at or not far above the so-called Planck scale of space and time (see Question 93)—many, many orders of magnitude below any region directly accessible to experiment. All we know for sure about that region is that spacetime itself must begin to participate in the quantum dance, resulting in "quantum foam."

According to the theory, different particles—of different mass, charge, spin, and so forth—result from different modes of vibration of the tiny strings, just as different notes on a violin result from different modes of vibration of its strings. The theory has charmed many fine minds by its mathematical elegance and its provocative hints about a possible unification of quantum theory and gravitation theory. It also has something unsettling to say about the number of dimensions in the universe—perhaps ten or eleven instead of the four (including time) of which we are aware. Yet it has so far had nothing to say about the world we can measure. There is, to date, no aspect of it that can be checked by experiment. Its allure, coupled with its remoteness, has led some physicists to argue that it is draining too much talent away from other avenues of physics. So there is a "sociology" of string theory, not just a mathematics of string theory.[†]

If the smallest dimension to which present-day experiments can probe is about 10^{-18} m and string theory is dealing with hypothetical

---

* A particle is accompanied by a swarm of virtual particles that extend over a spacetime region, but at the core of the swarm is a point particle.
† For a negative perspective (which has triggered quite a bit of debate), see Lee Smolin's *The Trouble with Physics* (New York: Houghton Mifflin, 2006).

phenomena at a dimension of some 10^{-35} m, almost a billion billion times smaller, how could we ever expect to confront the theory with experiment? There are several answers to this question. One answer is that it doesn't matter. If the theory gives a pleasing account of all that we know about the physical world and satisfies our craving for overarching simplicity in our description of nature, we will be inclined to accept it and believe it even if it predicts nothing new. Another answer is, "Who knows?" We shouldn't stop marching down some appealing path just because we aren't sure where it will lead. Many a breakthrough in physics has come from studies that at first seemed to have no relevance to the real world. And there is a third answer: that it is quite possible to conceive of testable predictions despite the gulf in time and distance scales that seems insurmountable. Suppose that someday some string theorist says, "I have just calculated that the mass of the muon should be 206.768 times the mass of the electron." You can be quite sure that string theory would be here to stay and that that theorist would receive an invitation to Stockholm.

100. What is the "measurement problem"? Much about quantum physics is strange, even eerie. Some writers speak of quantum weirdness. A photon that passes through two slits at once or a pair of atoms that remain entangled while possibly miles apart offends our sense of the way things ought to be. But where does that "sense" come from? As I discussed back at Question 6, it is the common sense derived from our everyday experiences in a world where classical physics rules and quantum phenomena remain below the surface, not directly noticed. We can speculate that nothing about quantum physics would seem weird to creatures who spend their daily lives in the midst of wave functions, virtual particles, superposition, quantum lumps, quantum jumps, and entanglement.

Physicists have come to accept that—until now, at least—all of the predictions of quantum physics are borne out by experiment. So, according to one line of thinking, there is no point in worrying about what quantum physics "means" or how it can be made to conform to our everyday ways of thinking. "Shut up and calculate," goes one aphorism (due to N. David Mermin, and often wrongly attributed to Richard Feynman*), meaning work out the consequences of quantum physics mathematically and compare prediction with experiment, whether or not it goes against your intuition or even your philosophical world view.

Yet beginning in the 1920s and continuing right up to the present, physicists *have* fretted about the meaning of quantum physics and how its predictions can be reconciled with a vision of a "real" physical world. Einstein didn't like probability and he didn't like entanglement. Many other physicists over the years have discussed and debated the "measurement problem," which focuses right on a central quantum weirdness. Almost from the earliest days of quantum theory, physicists were forced to think about how the evanescent wave function gets transformed into a specific here-and-now measurement. As far back as 1913 (when there were quantum jumps but not yet wave functions), Ernest Rutherford raised with Niels Bohr the question how does an electron in an excited state of motion know to what lower-energy state it is going to jump. As I remarked at Question 26, he could have added "and when it is going to jump." A dozen years later quantum theorists offered the concept of the "collapse of the wave function." Their idea was that an electron (or any quantum system) spreads through space as a wave, and when a measurement is performed, revealing a specific location or other property for the particle, the wave function "collapses." This is a way to think about the transition from probability to actuality.

The idea of a collapsing wave function is often associated with what is called the *Copenhagen interpretation* of quantum mechanics. One of its principal difficulties is that it draws a line between the quantum world, where wave functions and probabilities reign, and the classical world, where a counter clicks or a pointer moves or a neuron fires. But this is a

---

*Remarks about quantum weirdness are legion. Feynman reassured his students and readers that it is all right if they didn't understand quantum physics because he didn't understand it either. And Niels Bohr reportedly said that if your head doesn't swim when you think about quantum physics, you aren't understanding it.

fuzzy line. As Bohr himself recognized when he introduced the correspondence principle, the transition from the quantum to the classical world is gradual. There is no specific line that is crossed. One extreme view of the collapsing wave function idea is that human consciousness is involved in measurement. To me, at least, that makes little sense. Why not a microbe whose DNA is modified by a passing X-ray photon, or a bit of mica whose crystal structure is altered by a zooming muon? Those events, too, constitute "measurement."

Yet the "measurement problem" persists. What, exactly, is going on when the potentiality of a superposed quantum state is transformed into the actuality of a specific measurement? Back in the 1950s, a visionary graduate student at Princeton University, Hugh Everett, a student of John Wheeler, offered an answer to the question, an answer that has come to be known as the *many worlds interpretation* of quantum mechanics. The wave function doesn't collapse, he said—it divides. If a particle is in a superposed state of spin up and spin down, and a measurement is made revealing its spin to be pointing up, then, argued Everett, its spin remains both up and down—up in our world and down in some other world. It's not just the electron that spins. It is one's head. Imagining the subdividing and sub-subdividing of the wave function billions of times each second is more than most of us can manage. Yet there is a consistency to the Everett interpretation that has recommended it to even very conservative mathematical physicists. It is pretty clear that Niels Bohr didn't like it. John Wheeler was himself in a kind of superposed state between his own mentor Niels Bohr and his brilliant student Hugh Everett. It is fair to say that Everett "solved" the measurement problem—but at quite a cost. He died young (in 1982), not living to see an increased interest in his ideas more than a half-century after they were offered.*

Another Wheeler student, Wojciech Zurek, has become a leader in the study of "decoherence," which means, roughly, disentanglement or

---

*For a biography of Hugh Everett, see Peter Byrne's *The Many Worlds of Hugh Everett III: Multiple Universes, Mutual Assured Destruction, and the Meltdown of a Nuclear Family* (New York: Oxford University Press, 2010).

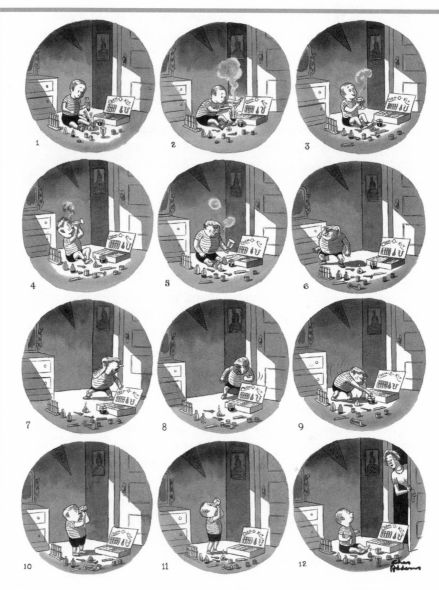

The mother's measurement determines her reality. Cartoon © Charles Addams. Used with permission of the Tee and Charles Addams Foundation.

desuperposition. Decoherence is a more sophisticated way of looking at the measurement problem. It replaces the idea of a collapsing wave function with the idea of a quantum system interacting with its "environment" (which usually means a larger, classical system) and thereby losing some or all of its quantum superposition features. Decoherence theory, like the many-worlds interpretation, does not make new predictions or in any way contradict "standard" quantum predictions. It provides a new and perhaps more satisfying way of looking at the world, especially the world at that fuzzy boundary between the classical and the quantum.

101. How come the quantum? I end the book with this favorite question of my old friend and mentor, John Wheeler.* No one knows the answer to the question or whether it even has an answer. As a young researcher in the 1930s, 1940s, and 1950s, Wheeler didn't question the "reason" for quantum mechanics. Like his mentor Niels Bohr, he accepted that it just reflects the way the world is—probability, uncertainty, waves and particles, superposition, entanglement, and all. Neither he nor any other researcher needed to know whether anything deeper lay beneath the theory. They just unleashed the theory as it existed to account with stunning success for myriad phenomena in the small-scale world.

But in his older years, Wheeler moved more toward the viewpoint of Einstein, coming to regard quantum mechanics as provisional, as a layer on top of a deeper—and perhaps simpler—core. He was troubled by the "measurement problem," that we achieve unambiguous results of measurement in the large-scale world that are built somehow on the ambiguities (the probabilities) of the small-scale world. Wheeler came to feel in his bones that some deeper theory, waiting to be discovered, would explain in a clear and rational way all the oddities of the quantum world, and

---

*Wheeler died in 2008 at the age of ninety-six after an astonishing career in which he made major contributions to nuclear physics, particle physics, quantum physics, and the physics of gravitation and relativity. He was also an important figure in the Manhattan Project and in the subsequent development of thermonuclear weapons.

would, in turn, explain the apparent fuzziness in the quantum-classical boundary.

Wheeler, who sometimes resorted to poetry, expressed his conviction this way:*

> *Behind it all*
> *is surely an idea so simple,*
> *so beautiful,*
> *so compelling that when—*
> *in a decade, a century,*
> *or a millennium—*
> *we grasp it,*
> *we will all say to each other,*
> *how could it have been otherwise?*
> *How could we have been so stupid*
> *for so long?*

Perhaps Wheeler's vision will be realized within the lifetime of readers of this book. I hope so. In the meantime, Question 101 remains unanswered.

---

* An unpublished poem, quoted by Kitty Ferguson in *Stephen Hawking: Quest for a Theory of Everything* (New York: Bantam, 1992), p. 21.

Appendix A

Table A.1 Leptons

Name	Symbol	Charge (unit: e)	Mass (unit: MeV)	Spin (unit: $\hbar$)	Antiparticle	Typical decay	Mean life
Flavor 1							
Electron	e	-1	0.511	½	e^+	Stable	
Electron neutrino	ν_e	0	Less than 2×10^{-6}	½	$\bar{\nu}_e$	Oscillates to other neutrinos	
Flavor 2							
Muon	μ	-1	105.7	½	μ^+	$\mu \rightarrow e + \nu_\mu + \bar{\nu}_e$	2.2×10^{-6} s
Muon neutrino	ν_μ	0	Less than 0.19	½	$\bar{\nu}_\mu$	Oscillates to other neutrinos	
Flavor 3							
Tau	τ	-1	1,776.8	½	τ^+	$\tau \rightarrow e + \nu_\tau + \bar{\nu}_e$	2.9×10^{-13} s
Tau neutrino	ν_τ	0	Less than 18	½	$\bar{\nu}_\tau$	Oscillates to other neutrinos	

Source: Particle Data Group of Lawrence Berkeley Laboratory, http://pdg.lbl.gov/.

Note: Half-life = 69.3 percent of mean life.

Table A.2 Quarks

Name	Symbol	Charge (unit: e)	Mass (unit: MeV)	Spin (unit: $\hbar$)	Baryon number	Antiparticle
Group 1						
Down	d	$-1/3$	4.5–5.5	$1/2$	$1/3$	$\bar{d}$
Up	u	$2/3$	1.8–3.0	$1/2$	$1/3$	$\bar{u}$
Group 2						
Strange	s	$-1/3$	90–100	$1/2$	$1/3$	$\bar{s}$
Charm	c	$2/3$	1,275	$1/2$	$1/3$	$\bar{c}$
Group 3						
Bottom	b	$-1/3$	4,200–4,600	$1/2$	$1/3$	$\bar{b}$
Top	t	$2/3$	174,000	$1/2$	$1/3$	$\bar{t}$

Source: Particle Data Group of Lawrence Berkeley Laboratory, http://pdg.lbl.gov/.

Table A.3 Some Important Bosons

Name	Symbol	Charge (unit: e)	Mass (unit: MeV)	Spin (unit: $\hbar$)	Antiparticle	Force that is carried
Graviton (hypothetical, never observed)	—	0	0	2	Self	Gravitational
W	W^+	1	80,400	1	W^-	Weak
Z	Z^0	0	91,190	1	Self	Weak
Photon	γ	0	0	1	Self	Electromagnetic
Gluon (set of 8 particles)	g	0 (but 3 "color charges")	0	1	Self	Strong
Higgs	H	0	125,000	0	Self	—

Source for the force carriers: Particle Data Group of Lawrence Berkeley Laboratory, http://pdg.lbl.gov/.
Note: The Higgs is not a force carrier.

Table A.4 Some Composite Particles

Name	Symbol	Charge (unit: e)	Mass (unit: MeV)	Quark composition	Spin (unit: $\hbar$)	Typical decay	Mean life
Baryons (which are fermions)							
Proton	p	1	938.3	uud	½	None known	More than 10^{29} years
Neutron	n	0	939.6	ddu	½	$n \rightarrow p + e + \bar{\nu}_e$	880 s
Lambda	Λ	0	1,116	uds	½	$\Lambda \rightarrow p + \pi^-$	2.6×10^{-10} s
Sigma	Σ	1, 0, −1	1,189 (+ & −) 1,193 (0)	uus (+), dds (−) uds (0)	½	$\Sigma^+ \rightarrow n + \pi^+$ $\Sigma^0 \rightarrow \Lambda + \gamma$	0.80×10^{-10} s (+ & −) 7×10^{-20} s (0)
Omega	Ω	−1	1,672	sss	3/2	$\Omega \rightarrow \Lambda + K^-$	0.82×10^{-10} s
Mesons (which are bosons)							
Pion	π	1, 0, −1	139.6 (+ & −) 135.0 (0)	$u\bar{d}$ (+), $d\bar{u}$ (−) $u\bar{u}$ & $d\bar{d}$ (0)	0	$\pi^+ \rightarrow \mu^+ + \nu_\mu$ $\pi^0 \rightarrow 2\gamma$	2.6×10^{-8} s (+ & −) 8×10^{-17} s (0)
Eta	η	0	548	$u\bar{u}$ & $d\bar{d}$	0	$\eta \rightarrow \pi^+ + \pi^0 + \pi^-$	5.6×10^{-19} s
Kaon	K	1, 0, −1	494 (+ & −) 498 (0)	$u\bar{s}$ (+), $\bar{u}s$ (−) $d\bar{s}$ & $\bar{d}s$ (0)	0	$K^- \rightarrow \mu^- + \bar{\nu}_\mu$ $K^0 \rightarrow \pi^+ + \pi^-$	1.24×10^{-8} s (+ & −) 0.89×10^{-10} s (0)

Source: Particle Data Group of Lawrence Berkeley Laboratory, http://pdg.lbl.gov/.
Note: Half-life = 69.3 percent of mean life.

Appendix B

Table B.1 **Large and Small Multipliers**

Factor	Name	Symbol	Factor	Name	Symbol
One hundred 10^2	Hecto	h	One hundredth 10^{-2}	Centi	c
One thousand 10^3	Kilo	k	One thousandth 10^{-3}	Milli	m
One million 10^6	Mega	M	One millionth 10^{-6}	Micro	μ
One billiion 10^9	Giga	G	One billionth 10^{-9}	Nano	n
One trillion 10^{12}	Tera	T	One trillionth 10^{-12}	Pico	p
One quadrillion 10^{15}	Peta	P	One quadrillionth 10^{-15}	Femto	f
One quintillion 10^{18}	Exa	E	One quintillionth 10^{-18}	Atto	a

Table B.2 How Big

Physical quantity	Typical magnitude in the quantum world
Length	Size of atom, about 10^{-10} m (0.1 nanometer, or 0.1 nm) Size of proton, about 10^{-15} m (1 femtometer, or 1 fm) "Planck length," about 10^{-35} m
Time	Time for a particle to cross a nucleus, about 10^{-23} s Typical half life of "long-lived" particle, about 10^{-10} s
Speed	Speed of light (c), 3×10^8 m/s Speed of electron in atom, about $0.01c$ to $0.1c$ Speed of particle in accelerator, very close to c
Mass	Mass of two electrons, about 1 million eV, or 1 MeV Mass of proton, about 1 billion eV, or 1 GeV (1 GeV of mass energy $= 1.78 \times 10^{-27}$ kg)
Energy	Air molecule at room temperature less than 1 eV Photon of green light, about 2 eV Electron in cathode ray tube more than 10^3 eV (1 keV) Proton in largest accelerator, 7×10^{12} eV (7 TeV)
Electric charge	Quantum unit of charge (e), 1.6×10^{-19} coulombs (1 coulomb illuminates a tiny 1-W night light for 1 second)
Spin	Quantum unit of spin ($\hbar$), about 10^{-34} kg $\times$ m $\times$ m/s Spin of photon, $\hbar$ Spin of electron or quark or proton, ($1/2$) $\hbar$

Acknowledgments

I am especially grateful to three friends for their help in making this book a reality: Diane Goldstein, for proposing the question-and-answer format and encouraging its execution; Jonas Schultz, for his careful reading of the manuscript and his many valuable suggestions; and Paul Hewitt, for his pedagogical insights and his many drawings that help bring the ideas in this book to life. I am indebted also to David Mermin for his astonishingly thorough review and for the insights into quantum physics that he shared with me.

Working with Michael Fisher and his team at Harvard University Press, including Anne Zarrella, has been the greatest pleasure. To Barb Goodhouse and Carrie Nelkin at the Westchester Book Group I am grateful for the firm but gentle—and effective—hands they applied to the work.

Rewards without end have been brought my way by my wife Joanne, my seven children, my fourteen grandchildren, and my innumerable wonderful students. It is to them I dedicate this book.

Index

Page numbers followed by the letter n refer to footnotes. Those followed by the letter t refer to tables. Italicized page numbers refer to illustrations.

Brownian motion, 9
bubble chamber, 214–215
bugle, quantized tones in, 54
Byrne, Peter, 261n
byte, 254n

carbon-12, 84
carbon-14, 84
    beta decay of, 90–91
    dating, 84
    half-life of, 91
cathode ray tube, 15, 207
cathode rays, 15
cavity radiation, 23–26
    Bose's derivation, 134
CERN, 125, 250
    and quark-gluon plasma, 242n
    See also Large Hadron Collider
certainty
    in quantum world, 68–69
Chadwick, James, 19, 80
chain reaction, 97–98
change
    in scattering and decay, 152–153
charge, 44–46
    conservation of, 45, 148–149, 152,
        155, 156t
    quantization of, 5, 45
    as a scalar quantity, 45
charge conjugation, 158–159
charm quark, 122
classical physics defined, 6
classical-quantum boundary, 7–8
closed shells
    in atoms, 80
    in nuclei, 80–81
cloud chamber, 212
cobalt-60, 164–165
"cold, dark matter," 228

collapse of wave function, 260
collective model, 20, 82
collector (of transistor), 227
color, 122–123
    conservation of, 152, 156–157
    definition of, 119
    of quarks, 124, *147*
color charge, 157
colorless particles, 122, 157
common sense vs. quantum sense, 1,
    14, 249
composite particles, 123–126
    sizes of, 125
Compton, Arthur, 134
Compton effect, 134
    and particle nature of light, 174
Condon, Edward, 63
conduction band, 219–220
    in quantum dots, 239
conductivity, 222
conductor, 218–220
confinement, 50–51
conservation
    as a "big idea," 2
    definition of, 169
    related to symmetry and invariance,
        169–172
conservation laws, 151–152
    absolute, 154–157
    classical, 155
    partial, 160–163
    See also individual laws
constant of proportionality, 41
continuum, appearance of, 54
Cooper, Leon, 233–234
Cooper pairs, 234–235, 237
Copenhagen interpretation, 260
copper, as excellent conductor, 219–220
Cornell, Eric, 132, 133n
correspondence principle, 7–9, 179, 261

cosmic rays, 39, 102, 208
  primary, 102n
Coulomb, Charles Augustin, 45
coupling constant(s), 45n, 245
Cowan, Clyde, Jr., 102, 105
CP invariance, 166–167
creation. *See* annihilation and creation
Cronin, James, 166–167
Curie, Marie, 88
"curviness" of wave function, 191, *192*
cyclotron, 208–209

Davis, Raymond, 91n, 111, 112, 115
Davisson, Clinton, 174, 175
de Broglie, Prince Louis Victor,
  174–175, 177–179, 190n
de Broglie equation, 174–177, 195
decay
  changes in, 152–153
  one-particle product prohibited, 149
decoherence, 262, 264
depletion zone, 223
detectors, 211–215
deuteron, stability of, 92–93
di-neutron, absence of, 93
di-proton, absence of, 87, 93
diffraction, 182–183
  definition of, 173, 182
  of electrons, 174, *175*
  in two-slit experiment, 184
diffraction grating, 70
dimensions in universe, 258
diodes, 223–224
  uses of, 224–225
Dirac, Paul, 136, 190n
  and magnetic monopoles, 245
dissociation
  definition of, 240n
  of matter, 240–242

distance
  scales of, 34–35
  smallest probed, 34
donors and acceptors, 221–222
doping, 221
doublets, 161
downhill rule, 65–66, 151, 155
dt reaction, 98

$E = mc^2$. *See* mass-energy equivalence
eigenfunction, 193n
Einstein, Albert, 28, 190n
  attitude toward quantum physics, 62,
    67, 185, 206, 248, 260
  and Bose-Einstein condensation, 132
  and bosons, 133–135
  and God, 67n
  and mass-energy equivalence, 41, *42*
  and Nobel Prize, 28, 30
  and the photon, 27
  and Planck, 29n
  quotation, 68
  and stimulated emission, 215–216
  See also EPR
Einstein-Marić, Mileva, 28
electric charge. *See* charge
electromagnetic force
  as exchange force, 36
  range of, 35
electromagnetic interaction, 144–147
electromagnetism, classical, 73
electron neutrino
  mass limit of, 109
  properties of, 268t
electron volt, 43
  definition of, 22n
electron(s)
  discovery of, 15–16
  identity of, 116–117, 127

symmetry (*continued*)
  related to invariance and
    conservation, 169–172
symmetry principles, partial, 163–167
synchrotrons, 209–210

tau, 108
  discovery of, 103–104
  properties of, 268t
tau neutrino, 108
  properties of, 268t
TCP theorem, 157–160
  applied to pion decay, 159–160
telephone dialer, 202
Teller, Edward, 218n
  relations with Bethe, 232n
  and solar fusion, 232
temperature
  of black hole, 230
  in Sun's center, 231
Teresi, Dick, 256n
Tevatron, 125, *126*
  energy of, 43
  and search for Higgs, 257
thermonuclear reactions, 100
Thomson, George, 15n, 174, 175
Thomson, J. J., 15–16, 101
three-prong vertices, 141
time, scales of, 40
time-reversal invariance, 166–167
  defined, 158
Tomonaga, Sin-Itiro, 137
top quark, 122
Townes, Charles, 217
transistors, 226–227
  n-p-n, *226*, 227
  p-n-p, 227
transmutation, 88
triode, 227

triplets, 161
triton, decay of, 108–109
Tukey, John, 253n
tunneling, 89, 186–188
  in alpha decay, 64
  and Josephson junctions, 237
  and solar fusion, 100, 231–233
two-slit experiment, 183–186

Uhlenbeck, George, 47–48, 60, 190n
uncertainty principle, 197–199
  as a "big idea," 2
  classical, 199–202
  classical *vs.* quantum, 202
  position-momentum form of, 198,
    201
  time-energy form of, 115–116,
    198–199
  and virtual particles, 146
  and waves, 199–202
unified model, 20, 82
unions in physics, 73
universal Fermi interaction, 145
ununoctium, 83n
uranium
  atom, 79
  isotopes of, 84
  isotopic differences, 97

vacuum tubes, 223n, 227n
valence band, 219–220
  in quantum dots, 239
valence electrons, 219–220
vector quantity, 47
vertices
  four-prong, 144, *145*
  nature of, 144
  three-prong, 141, 145